T0280318

Nonlinear Pinning Control of Complex Dynamical Networks

Nonlinear Pinning Control of Complex Dynamical Networks

Analysis and Applications

Edgar N. Sanchez

Carlos J. Vega

Oscar J. Suarez

Guanrong Chen

CRC Press
Taylor & Francis Group
Boca Raton London New York

CRC Press is an imprint of the
Taylor & Francis Group, an **informa** business

MATLAB® is a registered trademark of The MathWorks, Inc. For product information, please contact: The MathWorks, Inc. 3 Apple Hill Drive Natick, MA 01760-2098 USA Tel: 508-647-7000 Fax: 508-647-7001 Email: info@mathworks.com Web: www.mathworks.com

First edition published 2021
by CRC Press
6000 Broken Sound Parkway NW, Suite 300, Boca Raton, FL 33487-2742

and by CRC Press
2 Park Square, Milton Park, Abingdon, Oxon, OX14 4RN

© 2021 Edgar N. Sanchez, Carlos J. Vega, Oscar J. Suarez and Guanrong Chen

CRC Press is an imprint of Taylor & Francis Group, LLC

ISBN: 978-1-032-02087-7 (hbk)
ISBN: 978-1-032-02093-8 (pbk)
ISBN: 978-1-003-18180-4 (ebk)

Typeset in CMR10
by KnowledgeWorks Global Ltd.

To my wife: María de Lourdes,

our sons: Zulia Mayari, Ana María and Edgar Camilo,

and our grandsons: Esteban, Santiago and Amelia

Edgar N. Sanchez

To my parents: José Benjamín and Magali,

To my siblings: José Benjamín, Aura Magaly and Oscar Mauricio,

and my love: Laura Milena

Carlos J. Vega

To my parents: Oscar Enrique and Ana Rosa,

To my sister: Sara Juliana,

and my love: Marta Alejandra

Oscar J. Suarez

To my wife: Helen Q. Chen,

and our daughters: Julie and Leslie

Guanrong Chen

Contents

List of Figures

List of Tables

Preface

The study of various complex dynamical networks is currently pervading all kinds of sciences, such as physics, biology, and social sciences; its impact is significant for science and technology in different fields. Interest in complex networks has been increasing since the time of Euler in the eighteenth-century to recent studies. Different topologies and graphic characteristics in complex networks are described by Erdös and Rényi (ER) random graph model, Watts and Strogatz (WS) small-world model and Barabási and Albert (BA) scale-free model. Concerning the special structures of complex networks, a simple and effective control strategy named pinning control is spectacular, which was executed by applying local control to a small fraction of network nodes. Recent research results propose many quantitative measurements of complex networks, where three concepts play a key role: degree distribution, clustering coefficient, and average path length.

Considering the above advancements, the present book presents two nonlinear control strategies for complex dynamical networks. First, sliding-mode control is used, and then the inverse optimal control approach is employed. For both cases, model-based is considered in Chapters 3 and 5; then, in Chapters 4 and 6, both are based on determining a model for the unknown system using a recurrent neural network.

The book is organized in four sections. The first section covers analysis of the mathematical preliminaries, with a brief review for complex networks, and the pinning methodology. Additionally, sliding-mode control and inverse optimal control are introduced. Neural network structures are also discussed along with a description of the high-order ones. The second section presents the analysis and simulation results for sliding-mode control with identical or non-identical nodes. The third section describes analysis and simulation results for inverse optimal control with identical or non-identical nodes. Finally, the last section presents applications of these schemes, which are performed using gene regulatory networks and microgrids as examples.

The book summarizes research results obtained during the last four years at the Automatic Control Systems Laboratory of the CINVESTAV-IPN, with its name in Spanish: Center for Research and Advanced Studios of the National Polytechnic Institute, Guadalajara Campus.

Guadalajara, Mexico
January 2021

MATLAB$^{\circledR}$ is a registered trademark of The MathWorks, Inc. For product information, please contact:

The MathWorks, Inc. 3 Apple Hill Drive Natick, MA 01760-2098 USA Tel: 508-647-7000 Fax: 508-647-7001 Email: info@mathworks.com Web: www.mathworks.com

Acknowledgments

The authors thank the *National Council for Science and Technology* (CONACyT for its name in Spanish, Consejo Nacional de Ciencia y Tecnología), Mexico, for financial support under Project No. 257200, which allowed them to develop the research reported in this book.

The authors also thank CINVESTAV-IPN (Center for Research and Advanced Studies of the National Polytechnic Institute), Mexico for providing them with the outstanding environment and resources to develop the research, from 2016 to date.

Additionally, the authors are very grateful to Esteban A. Hernandez-Vargas, professor at the Universidad Nacional Autónoma de México, Queretaro, México; Ana E. González-Santiago, professor at the Universidad de Guadalajara, Biomedical Sciences Department, Tonalá, México; Otoniel Rodríguez-Jorge, professor at the Universidad Autónoma del Estado de Morelos, Biochemistry and Molecular Biology Department, Cuernavaca, México, and Larbi Djilali, professor at the Universidad Autónoma del Carmen, Ciudad del Carmen, México, for their support which assisted the authors in different ways and made this book possible.

Authors

Edgar N. Sanchez was born in 1949 in Sardinata, Colombia, South America. He obtained a BSEE, with a major in power systems, from Universidad Industrial de Santander (UIS), Bucaramanga, Colombia in 1971, an MSEE with a major in automatic control, from CINVESTAV-IPN (Center for Research and Advanced Studies of the National Polytechnic Institute), Mexico City, Mexico, in 1974, and the Docteur Ingenieur degree in automatic control from Institut Nationale Polytechnique de Grenoble, France in 1980. He was granted a National Research Council Award as a research associate at NASA Langley Research Center, Hampton, Virginia, USA (January 1985 to March 1987). His research interests center on neural networks and fuzzy logic as applied to automatic control systems. Since January 1997, he has been with CINVESTAV-IPN, Guadalajara Campus, Mexico, as a professor of electrical engineering graduate programs. He is the advisor of 24 PhD thesis and 42 M.Sc thesis students. He is a member of the Mexican National Research System (promoted to the highest rank, III, in 2005), the Mexican Academy of Science and the Mexican Academy of Engineering. He has published 7 books, more than 300 technical papers in international journals and conferences, and has served as associate editor and reviewer for different international journals and conferences. He has also been a member of many international IEEE and IFAC conference IPCs.

Carlos J. Vega was born in Bucaramanga, Colombia. He received the B.Sc. degree in Electronic Engineering and M.Sc. degree in Electronic Engineering both from the Universidad Industrial de Santander (UIS), Bucaramanga, Colombia, in 2014 and 2016, respectively. He received

D.Sc. in Electrical Engineering degree from the Center for Research and Advanced Studies of the National Polytechnic Institute (CINVESTAV-IPN), Guadalajara, Mexico in 2020. Actually, he is with School of Managment and Bussines, Universidad del Rosario, Bogota, Colombia. His research interests include complex networks, nonlinear control, inverse optimal control, neural networks, and power systems.

Oscar J. Suarez was born in Pamplona, Colombia. He received the B.Sc. in Electronic Engineering degree from the Universidad Pontificia Bolivariana (UPB), Bucaramanga, Colombia in 2011, the Master in Industrial Control degree from the Universidad de Pamplona, Pamplona, Colombia in 2015, and the D.Sc. in Electric Engineering degree from the Center for Research and Advanced Studies of the National Polytechnic Institute (CINVESTAV-IPN), Guadalajara, Mexico in 2019. Since January 2012, he has been a professor of engineering programs for undergraduate and graduate programs both in Colombia and Mexico. Currently, he works in the Mechatronics Engineering Department of the Universidad de Pamplona, Colombia, and is a Junior Research fellow of the Ministerio de Ciencia, Tecnología e Innovación (Minciencias) in Colombia. His research interests include complex networks, nonlinear control, neural networks, and systems biology.

Guanrong Chen received the M.Sc. degree in computer science from Sun Yat-sen University, Guangzhou, China, in 1981, and the Ph.D. degree in applied mathematics from Texas A&M University, College Station, TX, USA, in 1987. He was a tenured Full Professor with the University of Houston, Houston, TX, USA. He has been a Chair Professor and the Founding Director of the Centre for Chaos and Complex Networks, City University of Hong Kong, Hong Kong, since 2000. Dr. Chen is a member of the Academia of Europæa and a fellow of The World Academy of Sciences. He received the 2011 Euler Gold Medal in Russia and the conferred Honorary Doctorate by the Saint Petersburg State University, Russia, in 2011, and by the University of Le Havre, Nomandie, France, in 2014. He is a Highly Cited Researcher in Engineering as well as in Mathematics according to Thomson Reuters.

Part I

Analyses and Preliminaries

1

Introduction

1.1 Complex Dynamical Networks

Research on complex networks plays an outstanding role in sciences and engineering. A large set of interconnected nodes is typically referred to as a complex network, where each node is typically a nonlinear system. Many systems in the real-world have been modeled by various complex networks, such as social networks, biological networks, and technological networks [50]. For example, the brain consists of a large number of interconnected neurons. The organization and function of the cell is the effect of complex interactions set among genes, proteins, and other molecules. Social systems may be represented by graphs describing interactions among individuals. Ecosystems are composed of species whose interdependence can be mapped onto food webs. As one more example, large networked infrastructures, such as power grids and transportation networks, are fundamental to our modern society. In the study of complex networks, since the 1960s Erdos-Renyi (ER) [19] started the systematic analysis of random graphs; however, most complex networks in the real world are not completely random, i.e., they are not generated by a completely random process. The end of the 20th century was a turning point in network science research, due to two innovative research papers on complex networks. Specifically, Watts and Strogatz (WS) published an article about small-world networks [76], followed by Barabási and Albert

(BA) who introduced scale-free networks in [6]. These two articles revealed the most fundamental characteristic of the small-world property and the defining feature of the scale-free property of many complex networks. For dynamical networks, significant and interesting phenomena include the emergence of coordinated motions, synchronization, and consensus. These phenomena are present in nature and have been studied for a long time; one case which stimulated a great deal of interest is synchronization of coupled nonlinear dynamical systems [54, 73, 69]. Recently, another line of research, which has received considerable attention, is the study of synchronization in complex networks [40, 45]. On the other hand, consensus problems are also extensively studied, for example, analyzed in [52, 59, 53].

1.2   Pinning Control

Pinning control is used when a networked system is not synchronizable or if the synchronized state is not the desired one. For a large complex network, it is not possible to control every node; due to this an effective pinning control strategy can be used to guarantee synchronization of complex networks by applying local controllers on a small percentage of nodes [40, 72]. There is a large volume of literature on pinning control. Initially, Wang and Chen studied the stabilization of a scale-free dynamical network to its equilibrium via pinning control [72]. In [40], Li et al. derived stabilization of complex networks conditions for synchronization; they also established concatenation of pinning control for the whole network based on the concept of "virtual control," and analyzed the efficiency of selective pinning strategies. Interesting problems include complete synchronization [39, 79, 81], cluster synchronization [78, 41], selective strategies for pinning [4], control methodologies (adaptive control

[86], intermittent control [44], impulsive control [46], and so on), and time delays [80, 24].

1.3 Sliding-Mode Control

The sliding-mode control is a well-known discontinuous feedback control technique, which has been reviewed in several books and many journal articles. Theoretical analyses are presented in [18, 26, 70]. Relative simplicity for design, control of independent motion (maintaining sliding conditions), invariance to process dynamics, and external disturbances attenuation are main characteristics of sliding-mode control [26].

Recently, synchronization has become a subject for study with increasing attention. For most of chaos synchronization techniques, the master-slave or drive-response approach is used, where the basic idea is to design a controller to accomplish the slave system states to track the master system ones asymptotically. Different control techniques have been developed for synchronization of chaotic systems, especially sliding-mode control [56, 71, 82]. For finite-time synchronization of complex dynamical networks, theoretical analyses are reviewed in [34, 66, 85]. In [34], fixed-time synchronization of complex dynamical networks with non-identical nodes in the presence of bounded uncertainties and disturbances using sliding-mode control technique was developed. Furthermore, [66] applies nonsingular terminal sliding-mode control technique to realize the novel combination-combination synchronization between combination of two chaotic systems as drive system and combination of two chaotic systems as response system with unknown parameters in a finite time. Finally, [85] introduce the idea of combination synchronization into complex networks; based on sliding-mode control principle, the finite-time combination synchronization of four uncertain complex networks is investigated. In this regard, the main disadvantage to the above works is that all network nodes require a controller.

1.4   Optimal Nonlinear Control

Optimization is an important task, since maximization of benefits, or minimization of costs is always desirable. Optimal control is related to finding a control law for a given system such that a performance criterion is minimized. This criterion is usually formulated as a cost functional, which is a function of state and control variables, satisfying physical constraints. The optimal control problems can be solved considering two types of optimality conditions: Pontryagin necessary conditions ("Maximum principle") and Bellman sufficient conditions ("Dynamic Programming"). Dynamic programming was developed by R. E. Bellman in the late 1950s [7]. The dynamic programming approach to optimal control leads to a derivation of the partial differential Hamilton-Jacobi-Bellman equation (HJB) equation. The solution of the HJB equation, for nonlinear systems, is in general analytically intractable. The calculation of variations solution, on the other hand, only requires solving the well-known Euler-Lagrange equation which is more manageable than the HJB equation. However, these two solutions are not equivalent, since the Euler-Lagrange equation solves a trajectory optimization problem. The existence of a solution of the HJB equation is a sufficient condition for the optimal control solution [62].

At the present time, the HJB equation has rarely proved useful excluding the linear regulator problem. Considering that mathematical models of the systems are usually uncertain, it is desirable to obtain a robust optimal control scheme. Nevertheless, when the robust optimal control is considered, in which a disturbance term is involved in the system, the solution of the Hamilton-Jacobi-Isaacs (HJI) equation is required. Recently, adaptive dynamic programming has been extensively used to solve optimal control problems, obtaining adequate results for different real-world applications [49, 84]. These methods try to approximate the solution to the HJB or HJI equation using powerful tools, specially neural networks. In [42], synchronization of complex dynamical networks via optimal pinning control is achieved based on Pontryagin's minimum principle, with the roadblock that it is impossible to obtain an analytic solution; therefore, to obtain a solution, numerical methods are used. In [48], stabilization of complex networks via optimal control is developed based on the Ritz variation principle. As an alternative to bypass the solution of a partial differential HJB or HJI equation, the inverse-optimal control technique was developed, by first constructing a control Lyapunov function to achieve stabilization and then minimizing an associated cost functional, where this Lyapunov function solves the associated HJB or HJI equation [21, 36, 16, 62].

1.5 Artificial Neural Networks

The ultimate goal of control engineering is to implement automatic systems that could operate with increasing independence from human actions in an unstructured and uncertain environment. Such a system can be called autonomous or intelligent. It would need only to be presented with a goal and would achieve its objective by learning through continuous interaction with its environment by means of feedback about its behavior [57]. One class of models that has the capability to implement this learning is artificial neural networks. An artificial neural network is a massively parallelly distributed processor, inspired by biological neural networks, which can store experimental knowledge and have it available for use [30]. An artificial neural network consists of a finite number of neurons (structural elements), which are interconnected to each other. It has some similarities with the brain: knowledge is acquired through a learning process, and interneuron connectivity called synaptic weights are used to store this knowledge [57].

The mostly used neural network (NN) structures:

- Feedforward networks. In feedforward networks, the neurons are grouped into layers. Signals flow from the input to the output via unidirectional connections. The network exhibits a high degree of connectivity, contains one or more hidden layers of neurons, and the activation function of each network is smooth, generally a sigmoid function.

- Recurrent networks. In a recurrent neural network, the outputs of the neuron are fed back to the same neuron or neurons in the preceding layers. Signals flow in forward and backward directions.

The increasing use of NNs for modeling and control of nonlinear systems is mainly due to the following features, which make them particularly attractive [20]:

- NNs are universal approximators. It has been proven that any continuous nonlinear function can be approximated arbitrarily well over a compact set by a multilayer neural network consisting of one or more hidden layers [14].

- Learning and adaptation. The intelligence of NNs comes from their generalization ability with respect to unknown data. On-line adaptation of the weights is also possible.

NNs have become a well-established methodology as exemplified by their applications to identification and control of general nonlinear and complex systems [51]; the use of high-order neural networks for modeling and learning has recently increased [60]. Specifically, the problem of designing robust neural controllers for nonlinear systems with parametric uncertainties, unmodeled dynamics, and external disturbances, which guarantees stability and trajectory tracking, has received increasing attention lately. Using NNs, control algorithms can be developed to be robust in the presence of such events.

1.6   Gene Regulatory Networks

Gene regulatory networks play key roles in every process of life, including cell cycle, metabolism, signal transduction, cell communication, and cellular differentiation. These complex biological networks use large amounts of data, necessary for modeling, analyzing, and controlling. Mathematical and computational methods are very helpful approaches for constructing network models at molecule level to predict cell behavior under normal conditions or pathological ones. Network topology and interactions between nodes (representing molecules, proteins, genes, mRNA, others) and edges (establishing regulatory properties) describe the network dynamical behavior [8]. Different mathematical models have been developed for studying gene regulatory networks, which can be divided into four classes [15]: the first ones are logical models, which describe regulatory networks qualitatively, namely, Boolean networks [33, 2, 75], probabilistic Boolean networks [64, 65], Bayesian networks [22, 58], and Petri nets [11, 32]; the second ones are defined by continuous models such as ordinary differential equations [12, 68, 10], and the S-system formalism [35, 74]; the third ones are single-molecule level models [9, 17, 61], which account for interactions among molecules; and the last ones are hybrid models combining different formulations like discrete-time and continuous-time frameworks [1, 23]. The continuous-time approach consists in connecting a group of dependent variables to biochemical reaction kinetics. In this case, it is essential to assume that molecules have constant concentrations with respect to cellular compartments, in which their variations are continuous functions of time [12, 68, 10].

1.7 Microgrids

Microgrids provide an efficient solution to ensure better use of Distributed Energy Resource (DER), to improve operation and stability of the electrical grid. A microgrid is composed of different Distributed Generators (DGs) such as wind power systems, solar power systems, small gas turbines, among other electrical generators, storage devices, and loads [3]. To ensure flexible power-sharing and high-injected power quality, the DERs are linked to the microgrid using power electronic converters [83]. The hierarchical control scheme is extensively utilized for microgrids to ensure their adequate operation in two modes: grid-connected mode and islanded mode [27]. This scheme distributes the microgrid control objectives among three control levels as primary, secondary, and tertiary ones, to achieve the required performances [5].

In the hierarchical structure, the primary controller is used to achieve the tracking of voltages, currents, and powers, in which their desired values are obtained from the secondary layer [55]. This controller is installed at each DG in order to ensure power balance between power generation and the load [5]. The primary control objectives are adequately achieved by using conventional controllers as the droop method, which can be implemented without communication links between the microgrid subsystems [31, 37]. This technique has been used widely for microgrid applications in islanded mode [29, 67]. In [28], an improved droop control method is proposed, where two operations are included, which are error reduction operation and voltage recovery one. In [25], an enhanced droop control is investigated to improve transient droop performances. To improve the generated active and reactive power decoupling, enhanced droop controllers are developed in [13].

For island microgrid, the microgrid voltage amplitude and frequency should be maintained at predefined values in the presence of different operation conditions as explained in [63]. The secondary controller should set the generated active and reactive power for each DG regarding the voltage amplitude and frequency deviations. In [31], a voltage and current controllers and a modified droop method are used to design primary and secondary level control; the latter one is used to regulate the reactive power desired value. In [38], primary and secondary layers are developed such that the primary control based on the droop controller deals with power regulations, while the secondary one consists of a dynamics consensus algorithm, which is utilized to ensure the microgrid permanent connection to the main grid.

To ensure adequate sharing of the generated active and reactive power, using local controllers, and restore and maintain voltage amplitude and frequency of the islanded microgrid at the rated values, using the secondary controllers, are very difficult due to the complex dynamics and unexpected behaviors of loads as demonstrated in [29]. To solve the reactive power sharing problem, different works have been developed as in [47], where an adaptive voltage droop control is presented. This method provides a good solution for reactive power-sharing; however, nonlinear and unbalanced loads are not considered. Another method is developed in [77], where active power disturbances are used to identify the error of the reactive power sharing and eliminate it by using a slow integral term; however, this method affects power quality and can lead to instability.

1.8   Motivation

Taking into account the facts exposed above, the need to synthesize control algorithms for complex networks is obvious. These algorithms should be robust to external disturbances as well as parametric variations.

On the other hand, in most nonlinear control designs, it is usually assumed that the system model is previously known, as well as its parameters and disturbances. In practice, however, only part of this model is known. For this reason, identification remains an important topic, using for example neural identification.

Therefore, the major motivation for this book is to develop alternative methodologies, which allow the design of robust controllers for nonlinear systems with known or unknown dynamics.

Finally, it is highlighted that this book contains mathematical analysis, simulation examples, and real applications for the proposed schemes.

1.9   Book Structure

This book is organized as follows.

- In *Chapter 2*, mathematical preliminaries are introduced, including stability definitions, chaotic systems, complex dynamical networks, sliding-mode control, optimal control, and artificial neural networks.

- In *Chapter 3*, a novel control approach for trajectory tracking of complex networks is developed, based on sliding-mode pinning control technique. Two cases are presented. For the first case, the whole network tracks a reference for each one of the states; the second case uses the backstepping technique to track a desired trajectory for only one state. The illustrative example is composed of a network of 50 nodes; each node dynamics is a chaotic Chen's attractor.

- In *Chapter 4*, a novel control scheme to achieve output synchronization for uncertain complex networks with non-identical nodes is presented, based on neural sliding-mode pinning control technique. The control scheme is composed by an on-line identifier based on a recurrent high-order neural network, and a sliding-mode controller, where the former is used to build an on-line model for the unknown dynamics, and the latter to force the unknown node dynamics to track output synchronization.

- Then, in *Chapter 5*, a control approach to complex networks is developed, based on the technique of inverse optimal pinning control. This control law achieves stabilization of the tracking error dynamics. Chaotic systems are used to illustrate the applicability of the new methodology and the proposed controller verifying the effectiveness via simulations.

- After that, in *Chapter 6*, a control scheme for trajectory tracking on uncertain complex networks is proposed. A local adaptive strategy is developed based on a recurrent high-order neural network to identify the unknown system dynamics, and an inverse optimal controller applied for trajectory tracking. Furthermore, the development of a sampled-data pinning control scheme is presented for general complex networks to achieve trajectory tracking. Combining Lyapunov-Krasovskii approach and V-stability theory, a new criterion is derived to guarantee the stability of the error dynamics, thereby ensuring the whole complex network to achieve the control objective. The proposed control scheme is based on a neural identifier trained with an extended Kalman filter to model a pinned node, and the discrete-time inverse optimal controller is employed to ensure that the network tracks the reference state. The effectiveness of the control schemes is illustrated via simulations on an uncertain network of chaotic systems.

- *Chapter 7* presents pinning control for the p53-Mdm2 network. Theoretical and experimental studies show that the p53-Mdm2 network constitutes the core module of regulatory interactions activated by cellular stress induced by a variety of signaling pathways. Pinned nodes are selected on the basis of their importance level in a topological hierarchy, degree of connectivity in the network, and the biological role they perform. In this chapter, two cases are considered. For the first case, oscillatory pattern under gamma-radiation is recovered; for the second case, increased expression of p53 level is taken into account. For both cases, the control law is applied to p14ARF (pinned node), and overexpressed Mdm2-mediated p53 degradation condition is considered as initial behavior. As the main result of the proposed control technique, the two mentioned desired behaviors are obtained.

- *Chapter 8* includes an application of the discrete-time pinning neural control scheme in secondary control of microgrids. A distributed cooperative secondary controller is proposed to manage an islanded AC microgrid. Since the proposed secondary control is based on the pinning technique, all DGs are synchronized to the desired reference by means of a simple communication network, which leads to system reliability enhancement.

1.10   Notation

Throughout this book, the following notation will be used:

$k \in 0 \cup \mathbb{Z}^+$	Sampling step
$\lvert \bullet \rvert$	Absolute value
$\lVert \bullet \rVert$	Euclidian norm for vectors and matrices
$n \in \mathbb{R}$	Dimensional space of the system
$\mathbf{x} \in \mathbb{R}^n$	System state
$\mathbf{e} \in \mathbb{R}^n$	Tracking error
$N \in \mathbb{R}$	Number of nodes
$S\left(\bullet\right)$	Sigmoid function
$\chi \in \mathbb{R}^n$	Neural network state
$w_i^* \in \mathbb{R}^L$	ith neural network ideal weight vector
$L_i \in \mathbb{R}$	Number of high-order connections
$z_i \in \mathbb{R}^{L_i}$	High-order terms
$K_k \in \mathbb{R}^{L_i \times m}$	Kalman gain matrix
$P_k \in \mathbb{R}^{L_i \times L_i}$	Associated prediction error covariance matrix
$Q_k \in \mathbb{R}^{L_i \times L_i}$	Associated state noise covariance matrix
$r \in \mathbb{R}$	Number of blocks
$\mathbf{z}_i \in \mathbb{R}^{n_i}$	State transformation of the ith block
$\varepsilon \in \mathbb{R}^n$	Identification error

1.11   Acronyms

BA   Barabási and Albert

BFF   Block-Feedback Form

BCF   Block Controllable Form

CLF   Control Lyapunov Function

DC   Direct Current

DER   Distributed Energy Resource

DG   Distributed Generator

DNA   Deoxyribonucleic Acid

DSBs   Double-Strand Breaks

DT   Discrete-Time

ER   Erdos-Renyi

EKF   Extended Kalman Filter

HJB   Hamilton-Jacobi-Bellman

HJI   Hamilton-Jacobi-Isaacs

IOC   Inverse Optimal Control

ISS   Input-to-State Stable

KF   Kalman Filtering

MSE   Mean Square Error

NN   Neural Network

RHONN   Recurrent High-Order Neural Network

RNA   Ribonucleic Acid

SMC   Sliding-Mode Control

SPWM   Sinusoidal Pulse Width Modulation

TLEs   Transversal Lyapunov Exponents

WS   Watts and Strogatz

Bibliography

[1] J. Ahmad, G. Bernot, J. Comet, D. Lime, and O. Roux. Hybrid modelling and dynamical analysis of gene regulatory networks with delays. *ComPlexUs*, 3(4):231–251, 2006.

[2] T. Akutsu, S. Miyano, S. Kuhara. Identification of genetic networks from a small number of gene expression patterns under the Boolean network model. *Pacific Symposium on Biocomputing*, 17–28, 1999.

[3] A. Bidram, V. Nasirian, A. Davoudi, and F. Lewis. *Cooperative Synchronization in Distributed Microgrid Control*. Cham, Switzerland: Springer, 2017.

[4] A. Amani, M. Jalili, X. Yu, and L. Stone. Finding the most influential nodes in pinning controllability of complex networks. *IEEE Transactions on Circuits and Systems II: Express Briefs*, 64(6):685–689, 2017.

[5] M. Awal, H. Yu, H. Tu, S. Lukic, and I. Husain. Hierarchical control for virtual oscillator based grid-connected and islanded microgrids. *IEEE Transactions on Power Electronics*, 35(1):988–1001, 2019.

[6] A. Barabási and R. Albert. Emergence of scaling in random networks. *Science, American Association for the Advancement of Science*, 286(5439):509–512, 1999.

[7] R. Bellman. *Dynamic Programming*. Princeton, NJ, USA: Princeton University Press, 1957.

[8] H. Bolouri and E. Davidson. Modeling transcriptional regulatory networks. *BioEssays*, 24(12):1118–1129, 2002.

[9] L. Cai, N. Friedman, and X. Xie. Stochastic protein expression in individual cells at the single molecule level. *Nature*, 440(7082):358, 2006.

[10] J. Cao, X. Qi, and H. Zhao. Modeling gene regulation networks using ordinary differential equations. *Next Generation Microarray Bioinformatics*, 185–197, 2012.

[11] C. Chaouiya. Petri net modelling of biological networks. *Briefings in Bioinformatics*, 8(4):210–219, 2007.

[12] T. Chen, H. He, and G. Church. Modeling gene expression with differential equations. *Biocomputing'99*, 29–40, 1999.

[13] S. Chiang, C. Yen, and K. Chang. A multimodule parallelable series-connected PWM voltage regulator. *IEEE Transactions on Industrial Electronics*, 3(48):506–516, 2001.

[14] N. Cotter. The Stone-Weierstrass theorem and its application to neural networks. *IEEE Transactions on Neural Networks*, 1(4):290–295, 1990.

[15] H. De Jong. Modeling and simulation of genetic regulatory systems: A literature review. *Journal of Computational Biology*, 9(1):67–103, 2002.

[16] H. Deng and M. Krstic. Output-feedback stochastic nonlinear stabilization. *IEEE Transactions on Automatic Control*, 44(2):328–333, 1999.

[17] J. Elf, G. Li, and X. Xie. Probing transcription factor dynamics at the single-molecule level in a living cell. *Science*, 316(5828):1191–1194, 2007.

[18] S. Emelyanov. *Variable Structure Control Systems*. Nouka, Moscow-Russia, 1967.

[19] P. Erdős and A. Rényi. On the evolution of random graphs. *Institute of Mathematics Hungarian Academy of Sciences*, 5(1):17–60, 1960.

[20] R. Felix. Variable Structure Neural Control. *PhD Dissertation, Cinvestav, Unidad Guadalajara, Guadalajara, Jalisco, Mexico*, 2004.

[21] R. Freeman and P. Kokotovic. Inverse optimality in robust stabilization. *Journal on Control and Optimization*, 34(4):1365–1391, 1996.

[22] N. Friedman, M. Linial, I. Nachman and D. Pe'er. Using Bayesian networks to analyze expression data. *Journal of Computational Biology*, 7(3-4):601–620, 2000.

[23] J. Fromentin, D. Eveillard, and O. Roux. Hybrid modeling of biological networks: Mixing temporal and qualitative biological properties. *BMC Systems Biology*, 4(1):79, 2010.

[24] D. Gong, H. Zhang, Z. Wang, and B. Huang. Pinning synchronization for a general complex networks with multiple time-varying coupling delays. *Neural Processing Letters*, 35(3):221–231, 2012.

[25] J. Guerrero, L. de Vicma, M. Castilla, and J. Miret. A Wireless controller to enhance dynamic performance of parallel inverters in distributed generation system. *IEEE Transactions on Power Electronics*, 5(19):1205–1213, 2004.

[26] J. Guldner and V. I. Utkin. Tracking the gradient of artificial potential fields: Sliding mode control for mobile robots. *International Journal of Control*, 63(3):417–432, 1996.

[27] F. Guo, C. Wen, Y. Song. *Distributed Control and Optimization Technologies in Smart Grid Systems*. Boca Raton, FL, USA: CRC Press, 2017.

[28] H. Han, Y. Liu, Y. Sun, M. Su, and J. Guerrero. An improved droop control strategy for reactive power sharing in islanded microgrid. *IEEE Transactions on Power Electronics*, 6(30):3133–3141, 2015.

[29] Y. Han, H. Li, P. Shen, A. Ernane, A. Coelho, J. Guerrero. Review of active and reactive power sharing strategies in hierarchical controlled microgrids. *IEEE Transactions on Power Electronics*, 3(32):2427–2451, 2016.

[30] S. Haykin. *Kalman Filtering and Neural Networks*. New York, NY, USA: John Wiley & Sons, 2001.

[31] I. Sadeghkhani, M. Golshan, A. Mehrizi, and J. Guerrero. Low-voltage ride-through of a droop-based three-phase four-wire grid-connected microgrid. *IET Generation, Transmission & Distribution*, 8(12):1906–1914, 2018.

[32] G. Karlebach and R. Shamir. Modelling and analysis of gene regulatory networks. *Nature Reviews Molecular Cell Biology*, 9(10):770, 2008.

[33] S. Kauffman. Metabolic stability and epigenesis in randomly constructed genetic nets. *Journal of Theoretical Biology*, 22(3):437–467, 1996.

[34] A. Khanzadeh and M. Pourgholi. Fixed-time sliding mode controller design for synchronization of complex dynamical networks. *Nonlinear Dynamics*, 88(4):2637–2649, 2017.

[35] S. Kikuchi, D. Tominaga, M. Arita, K. Takahashi, and M. Tomita. Dynamic modeling of genetic networks using genetic algorithm and S-system. *Bioinformatics*, 19(5):643–650, 2003.

[36] M. Krstic and Z. H. Li. Inverse optimal design of input-to-state stabilizing nonlinear controllers. *IEEE Transactions on Automatic Control*, 43(3):336–350, 1998.

[37] L. Djilali, E. Sanchez, F. Ornelas-Tellez, A. Avalos, and M. Belkheiri. Improving microgrid low-voltage ride-through capacity using neural control. *IEEE Systems Journal*, 2019.

[38] W. Lee, T. Nguyen, H. Yoo, and H. Kim. Low-voltage ride-through operation of grid-connected microgrid using consensus-based distributed control. *Energies*, 11(11):2867, 2018.

[39] X. Li and G. Chen. Synchronization and desynchronization of complex dynamical networks: An engineering viewpoint. *IEEE Transactions on Circuits and Systems I: Fundamental Theory and Applications*, 50(11):1381–1390, 2003.

[40] X. Li, X. Wang, and G. Chen. Pinning a complex dynamical network to its equilibrium. *IEEE Transactions on Circuits and Systems I: Regular Papers*, 51(10): 2074—2087, 2004.

[41] L. Li and J. Cao. Cluster synchronization in an array of coupled stochastic delayed neural networks via pinning control. *Neurocomputing*, 74(5): 846–856, 2011.

[42] K. Li, W. Sun, M. Small, and X. Fu. Practical synchronization on complex dynamical networks via optimal pinning control. *Physical Review E*, 92(1):010903, 2015.

[43] P. Li, B. Wang, W. Lee, and D. Xu . Dynamic power conditioning method of microgrid via adaptive inverse control. *IEEE Transactions on Powers Delivery*, 2(30):906–913, 2015.

[44] X. Liu and T. Chen. Synchronization of complex networks via aperiodically intermittent pinning control. *IEEE Transactions on Automatic Control*, 60(12): 3316–3321, 2015.

[45] J. Lu and G. Chen. A time-varying complex dynamical network model and its controlled synchronization criteria. *IEEE Transactions on Automatic Control*, 50(6): 841–846, 2005.

[46] J. Lu, J. Kurths, J. Cao, N. Mahdavi, and C. Huang. Synchronization control for nonlinear stochastic dynamical networks: Pinning impulsive strategy. *IEEE Transactions on Neural Networks and Learning Systems*, 23(2): 285–292, 2012.

[47] H. Mahmood, D. Michaelson, and J. Jiang. Reactive power sharing in island microgrids using adaptive voltage droop control. *IEEE Transactions on Smart Grid*, 6(6): 3052–3060, 2015.

[48] G. Mei, X. Wu, D. Ning, and J. Lu. Finite-time stabilization of complex dynamical networks via optimal control. *Complexity*, 21(S1): 417–425, 2016.

[49] H. Modares, F.L. Lewis, and Z.P. Jiang. H_∞ tracking control of completely unknown continuous-time systems via off-policy reinforcement learning. *IEEE Transactions on Neural Networks and Learning Systems*, 26(10): 2550–2562, 2015.

[50] M. Newman. The structure and function of complex networks. *SIAM Review*, 45(2):167–256, 2003.

[51] M. Norgaard, O. Ravn, N. Poulsen, and L. Hansen. *Neural Networks for Modelling and Control of Dynamic Systems*. New York, USA: Springer Verlag, 2000.

[52] R. Olfati-Saber and R. Murray. Consensus problems in networks of agents with switching topology and time-delays. *IEEE Transactions on automatic control*, 49(9):1520–1533, 2004.

[53] A. Papachristodoulou, A. Jadbabaie, and U. Munz. Effects of delay in multi-agent consensus and oscillator synchronization. *IEEE Transactions on Automatic Control*, 55(6):1471–1477, 2010.

[54] L. Pecora and T. Carroll. Master stability functions for synchronized coupled systems. *Physical Review Letters*, 80(10):2109, 1998.

[55] N. Pogaku, M. Prodanovic, and T. Green. Modeling, analysis and testing of autonomous operation of an inverter-based microgrid. *IEEE Transactions on Power Electronics*, 22(2): 613–625, 2007.

[56] M. Pourmahmood, S. Khanmohammadi, and G. Alizadeh. Finite-time synchronization of two different chaotic systems with unknown parameters via sliding mode technique. *Applied Mathematical Modelling*, 35:3080–3091, 2011.

[57] A. Poznyak, E. Sanchez, and W. Yu. *Differential Neural Networks for Robust Nonlinear Control: Identification, State Estimation and Trajectory Tracking*. World Scientific, 2001.

[58] A. Rau, F. Jaffrézic, J. Foulley, and R. Doerge. An empirical Bayesian method for estimating biological networks from temporal microarray data. *Statistical Applications in Genetics and Molecular Biology*, 9(1),2010.

[59] W. Ren. On consensus algorithms for double-integrator dynamics. *IEEE Transactions on Automatic Control*, 53(6):1503–1509, 2008.

[60] E. N. Sanchez, A. Y. Alanís, and A. G. Loukianov. *Discrete-Time High Order Neural Control: Trained with Kalman Filtering*. Berlin, Germany: Springer, 2008.

[61] P. Selvin and T. Ha. *Single-Molecule Techniques*. New York, NY, USA: Cold Spring Harbor, 2008.

[62] R. Sepulchre, M. Jankovic, and P. Kokotovic. *Constructive Nonlinear Control.* Berlin, Germany: Springer, 2012.

[63] X. Shen, H. Wang, J. Li, Q. Su, and L. Gao. Distributed secondary voltage control of islanded microgrids based on RBF-neural network sliding-mode technique. *IEEE Access,* 7:65616–65623, 2019.

[64] I. Shmulevich, E. Dougherty, and W. Zhang. From Boolean to probabilistic Boolean networks as models of genetic regulatory networks. *Proceedings of the IEEE,* 90(11):1778–1792, 2002.

[65] I. Shmulevich, E. Dougherty, S. Kim, and W. Zhang. Probabilistic Boolean networks: A rule-based uncertainty model for gene regulatory networks. *Bioinformatics,* 18(2):261–274, 2002.

[66] J. Sun, Y. Shen, X. Wang, and J. Chen. Finite-time combination-combination synchronization of four different chaotic systems with unknown parameters via sliding mode control. *Nonlinear Dynamics,* 76(1):383–397, 2014.

[67] Y. Sun, X. Hou, J. Yang, H. Han, M. Su, and J. Guerrero. New perspectives on droop control in AC microgrid. *IEEE Transactions on Industrial Electronics,* 7(64):5741–5745, 2017.

[68] Z. Szallasi, J. Stelling, and V. Periwal. *System Modeling in Cellular Biology: From Concepts to Nuts and Bolts.* London, UK: The MIT Press, 2006.

[69] S. Tuna. Conditions for synchronizability in arrays of coupled linear systems. *IEEE Transactions on Automatic Control,* 54(10):2416–2420, 2009.

[70] V. I. Utkin. *Sliding Modes and their Application in Variable Structure Systems.* Moscow, Russia: Mir Publishers, 1978.

[71] S. Vaidyanathan and S. Sampath. Global chaos synchronization of hyperchaotic Lorenz systems by sliding mode control. *Advances in Digital Image Processing and Information Technology*, 156–164, 2011.

[72] X. Wang and G. Chen. Pinning control of scale-free dynamical networks. *Physica A: Statistical Mechanics and Its Applications*, 310(3):521–531, 2002.

[73] X. Wang and G. Chen. Complex networks: Small-world, scale-free and beyond. *IEEE Circuits and Systems Magazine*, 3(1):6–20, 2003.

[74] H. Wang, L. Qian, and E. Dougherty. Inference of gene regulatory networks using S-system: A unified approach. *IET Systems Biology*, 4(2):145–156, 2010.

[75] R. Wang, A. Saadatpour, and R. Albert. Boolean modeling in systems biology: An overview of methodology and applications. *Physical Biology*, 9(5):055001, 2012.

[76] D. Watt and S. Strogatz. Collective dynamics of small-world networks. *Nature*, 393(6684):440–442, 1998.

[77] J. He and Y. Li. An enhanced microgrid load demand sharing strategy. *IEEE Transactions on Power Electronics*, 9(27):3984–3995, 2012.

[78] W. Wu, W. Zhou, and T. Chen. Cluster synchronization of linearly coupled complex networks under pinning control. *IEEE Transactions on Circuits and Systems I: Regular Papers*, 56(4):829–839, 2009.

[79] J. Xiang and G. Chen. On the V-stability of complex dynamical networks. *Automatica*, 43(6):1049—1057, 2007.

[80] L. Xiang, Z. Chen, Z. Liu, F. Chen, and Z. Yuan. Pinning control of complex dynamical networks with heterogeneous delays. *Computers & Mathematics with Applications*, 56(5):1423–1433, 2008.

[81] J. Xiang and G. Chen. Analysis of pinning-controlled networks: A renormalization approach. *IEEE Transactions on Automatic Control*, 54(8):1869–1875, 2009.

[82] H. Yau. Design of adaptive sliding mode controller for chaos synchronization with uncertainties. *Chaos, Solitons and Fractals*, 22:341–347, 2004.

[83] M. Yazdanian and A. Mehrizi. Distributed control techniques in microgrids. *IEEE Transactions on Smart Grid*, 5(6):2901–2909, 2014.

[84] H. Zhang, H. Jiang, C. Luo, and G. Xiao. Discrete-time nonzero-sum games for multiplayer using policy-iteration-based adaptive dynamic programming algorithms. *IEEE Transactions on Cybernetics*, 47(10):3331–3340,2017.

[85] M. Zhang, M. Xu, and M. Han. Finite-time combination synchronization of uncertain complex networks by sliding mode control. *Information, Cybernetics and Computational Social Systems*, 406–411, 2017.

[86] J. Zhou, J. Lu, and J. Lu. Adaptive synchronization of an uncertain complex dynamical network. *IEEE Transactions on Automatic Control*, 51(4):652–656, 2006.

2

Preliminaries

This chapter presents mathematical preliminaries, with a brief review for complex networks, and the pinning methodology. Additionally, sliding-mode control and inverse optimal control are introduced. Neural network structures are also discussed along with a description of the high-order ones.

2.1 Nonlinear Systems Stability

Consider a general non-autonomous or time-variant nonlinear system

$$\dot{\mathbf{x}}(t) = f(\mathbf{x}, t), \quad \mathbf{x}(t) \in \mathbb{R}^n \tag{2.1}$$

where $f : [0, \infty) \times D \to \mathbb{R}^n$ is a piecewise continuous function in $\mathbf{x}$ and $D \subset \mathbb{R}^n$ is a domain that contains the origin $\mathbf{x} = 0$.

Definition 2.1. The origin is an equilibrium point of (2.1) if

$$f(t, 0) = 0 \quad \forall t \geq 0$$

Lyapunov stability theory concerns various stabilities of this equilibrium of the system.

Definition 2.2 (Stability in the sense of Lyapunov)**.** The equilibrium $\mathbf{x} = 0$

of system (2.1) is said to be *stable in the sense of Lyapunov*, if for any $\varepsilon > 0$, and any initial time $t_0 \geq 0$, there is $\delta = \delta(\varepsilon, t_0) > 0$, such that

$$\| \mathbf{x}(t_0) \| < \delta \Rightarrow \| \mathbf{x}(t) \| < \varepsilon, \quad \forall t \geq t_0$$

where the constant δ is in general dependent on the initial time t_0.

Definition 2.3 (Lyapunov stability). The equilibrium $\mathbf{x} = 0$ of the system (2.1) is

- *Uniformly stable* if and only if there exist a class $\mathcal{K}$ function $\gamma(\cdot)$, and a positive constant c, independent of t_0, such that

$$\| \mathbf{x}(t) \| \leq \gamma(\| \mathbf{x}(t_0) \|), \quad \forall \quad t \geq t_0 \geq 0, \quad \| \mathbf{x}(t_0) \| \leq c$$

- *Uniformly asymptotically stable*, if and only if there exists a class $\mathcal{KL}$ function $\beta(\cdot, \cdot)$ and a positive constant c, independent of t_0, such that

$$\| \mathbf{x}(t) \| \leq \beta(\| \mathbf{x}(t_0) \|, t - t_0), \quad \forall t \geq t_0 \geq 0, \quad \forall \quad \| \mathbf{x}(t_0) \| \leq c \quad (2.2)$$

- *Globally uniformly asymptotically stable* if and only if (2.2) is satisfied for any initial state $\mathbf{x}(t_0)$.

- *Exponentially stable* if there exists positive constants c, k and λ such that

$$\beta(r, s) = k \| \mathbf{x}(t_0) \| e^{-\lambda(t - t_0)}, \quad \forall \quad \| \mathbf{x}(t_0) \| \leq c$$

- *Globally exponentially stable* if the foregoing inequality holds $\forall \quad \mathbf{x}(t_0)$.

For the autonomous nonlinear system

$$\dot{\mathbf{x}} = f(\mathbf{x}), \quad \mathbf{x}(t_0) = x_0 \quad (2.3)$$

where $\mathbf{x} \in \mathbb{R}^n$, $f : D \to \mathbb{R}^n$ is a continuously differentiable function, the following criterion of stability is called the first (or indirect) method of Lyapunov. Let

$$Df(\mathbf{x}) = \left[\frac{\partial f}{\partial \mathbf{x}}\right]_{\mathbf{x}=0}$$

be the system Jacobian matrix evaluated at the zero equilibrium.

Theorem 2.1 (First method of Lyapunov, for autonomous systems [10]). *If all the eigenvalues of $Df(\mathbf{x})$ have a negative real part, then the system (2.3) is asymptotically stable about $\mathbf{x} = 0$.*

Theorem 2.2 (Second method of Lyapunov, for non-autonomous systems [10]). *The system (2.1) is globally (over the entire domain D), uniformly (with respect to the initial time over the entire time interval $[t_0, \infty)$), and asymptotically stable about its zero equilibrium, if there exists a scalar-valued function, $V(\mathbf{x}, t)$, defined on $D \times [t_0, \infty)$, and three functions $\alpha(\cdot)$, $\beta(\cdot)$, $\gamma(\cdot)$ $\in \mathcal{K}$, such that*

1. *$V(0, t_0) = 0$;*

2. *$V(\mathbf{x}, t) > 0$ for all $\mathbf{x} \neq 0$ in D and all $t \geq t_0$;*

3. *$\alpha(\|\mathbf{x}\|) \leq V(\mathbf{x}, t) \leq \beta(\|\mathbf{x}\|)$ for all $t \geq t_0$;*

4. *$\dot{V}(\mathbf{x}, t) \leq -\gamma(\|\mathbf{x}\|)$ for all $t \geq t_0$.*

In this theorem, the function V is called a *Lyapunov function*. The method of constructing a Lyapunov function for stability determination is called the *second (or direct) method of Lyapunov*.

Theorem 2.3 (Second method of Lyapunov, for autonomous systems [10]). *The system (2.3) is globally (over the entire domain D) and asymptotically stable about its zero equilibrium, if there exists a scalar-valued function, $V(\mathbf{x})$, defined on D, such that*

1. *$V(0) = 0$;*

2. *$V(\mathbf{x}) > 0$ for all $\mathbf{x} \neq 0$ in D;*

3. $\dot{V}(\mathbf{x}) < 0$ *for all* $\mathbf{x} \neq 0$ *in* D.

Note that if condition (3) in the above theorem is replaced by

$$\dot{V}(\mathbf{x}) \leq 0 \text{ for all } \mathbf{x} \neq 0 \text{ in } D,$$

then the resulting stability is only in the sense of Lyapunov but may not be asymptotic.

Theorem 2.4 (LaSalle Theorem [10]). *Let* $\Omega \subset D$ *be a compact set that is positively invariant with respect to* (2.1). *Let* $V : \mathbb{R}^n \to \mathbb{R}$ *be a continuously differentiable function such that* $\dot{V}(\mathbf{x}) \leq 0$ *in* Ω. *Let* E *be the set of all points in* Ω *where* $\dot{V}(\mathbf{x}) = 0$. *Let* M *be the largest invariant set in* E. *Then, every solution starting in* Ω *approaches* M *as* $t \to \infty$.

Corollary 2.1 ([10]). *Let* $\mathbf{x} = 0$ *be an equilibrium point for* (2.1). *Let* $V : \mathbb{R}^n \to \mathbb{R}$ *be a continuously differentiable, positive definite function on a domain* D *containing the origin* $\mathbf{x} = 0$, *such that* $\dot{V}(\mathbf{x}) \leq 0$ *in* D. *Let* $S = \{\mathbf{x} \in \mathbb{R}^n \mid \dot{V}(\mathbf{x}) = 0\}$ *and suppose that no solution can stay permanently in* S, *other than the trivial solution* $\mathbf{x} \equiv 0$. *Then, the origin is asymptotically stable.*

Corollary 2.2 ([10]). *Let* $\mathbf{x} = 0$ *be an equilibrium point for* (2.1). *Let* $V : \mathbb{R}^n \to \mathbb{R}$ *be a continuously differentiable, radially unbounded, positive definite function such that* $\dot{V}(\mathbf{x}) \leq 0$ *for all* $\mathbf{x} \in \mathbb{R}^n$. *Let* $S = \{\mathbf{x} \in \mathbb{R}^n \mid \dot{V}(\mathbf{x}) = 0\}$ *and suppose that no solution can stay permanently in* S, *other than the trivial solution* $\mathbf{x} \equiv 0$. *Then, the origin is globally asymptotically stable.*

Lemma 2.1 (Barbalat's Lemma [10]). *Let* $v : \mathbb{R}^n \to \mathbb{R}$ *be a uniformly continuous function on* $[0, \infty)$. *Suppose that*

$$\lim_{t \to \infty} \int_0^t v(\tau) d\tau$$

exists and is finite. Then,

$$v \to 0 \text{ as } t \to \infty$$

Theorem 2.5 ([2]). *Suppose that there is a continuous function $V : D \to \mathbb{R}$, such that the following conditions hold:*

(i) V is positive definite.

(ii) There exist real numbers $c > 0$ and $\psi \in (0,1)$ and an open neighborhood $\nu \subseteq D$ of the origin, such that

$$\dot{V}(\mathbf{x}) + c(V(\mathbf{x}))^{\psi} \le 0, \quad \mathbf{x} \in \nu \backslash \{0\}. \tag{2.4}$$

Then, the origin is a finite-time-stable equilibrium of (2.3). Moreover, if $\mathcal{N}$ is an open neighborhood $\mathcal{N} \subseteq D$ and t_s is the settling-time function, then

$$t_s(\mathbf{x}) \le \frac{1}{c(1-\psi)} V(\mathbf{x})^{1-\psi}, \quad \mathbf{x} \in \mathcal{N}, \tag{2.5}$$

and t_s is continuous on $\mathcal{N}$. If in addition $D = \mathbb{R}^n$, V is proper, and $\dot{V}$ takes negative values on $\mathbb{R}^n \backslash \{0\}$, then the origin is a globally finite-time-stable equilibrium of (2.3).

Definition 2.4 (The Lie derivative [10]). Let $h : D \to \mathbb{R}$ be a scalar function and $f : D \to \mathbb{R}^n$ be a sufficiently smooth vector field in a domain $D \subset \mathbb{R}^n$. The Lie derivative of h with respect to f, or along f, written as $L_f h$, is defined by

$$L_f h(\mathbf{x}) = \frac{\partial h}{\partial \mathbf{x}} f(\mathbf{x}).$$

Theorem 2.6. *Let $D = \{\mathbf{x} \in \mathbb{R}^n \mid \|\mathbf{x}\| < r\}$ and suppose that $f(t,\mathbf{x})$ is piecewise continuous in t, locally Lipschitz in $\mathbf{x}$, uniformly in t, on $[0,\infty) \times D$. Let $V : [0,\infty) \times D \to \mathbb{R}$ be a continuously differentiable function such that*

$$W_1(\mathbf{x}) \le V(t,\mathbf{x}) \le W_2(\mathbf{x})$$
$$\dot{V}(t,\mathbf{x}) = \frac{\partial V}{\partial t} + \frac{\partial V}{\partial \mathbf{x}} f(t,\mathbf{x}) \le -W(\mathbf{x})$$

$\forall t \geq 0$, $\forall \mathbf{x} \in D$, *where* $W_1(\mathbf{x})$ *and* $W_2(\mathbf{x})$ *are continuous positive definite functions and* $W(\mathbf{x})$ *is a continuous positive semi-definite function on* D. *Let* $\rho < min_{\|\mathbf{x}\|=r} W_1(\mathbf{x})$. *Then, all the solutions of* $\dot{\mathbf{x}} = f(t, \mathbf{x})$ *with* $\mathbf{x}(t_0) \in \{\mathbf{x} \in B_r \mid W_2(\mathbf{x}) \leq \rho\}$ *are bounded and satisfy*

$$W(\mathbf{x}(t)) \to 0 \quad as \quad t \to \infty$$

Moreover, if the assumptions hold globally and $W_1(\mathbf{x})$ *is radially unbounded, then the statement is true for all* $\mathbf{x}(t_0) \in \mathbb{R}^n$.

Definition 2.5 (Input-to-state stability (ISS) [11]). Consider the system

$$\dot{\mathbf{x}} = f(t, \mathbf{x}, u) \tag{2.6}$$

where $f : [0, \infty)\mathbb{R}^n \times \mathbb{R}^m \to \mathbb{R}^n$ is piecewise continuous in t and locally Lipschitz in $\mathbf{x}$ and u. The input $u(t)$ is a piecewise continuous, bounded function of t, for all $t \geq 0$. The system (2.6) is said to be input-to-state stable if there exists a class $\mathcal{KL}$ function β and a class K function γ such that for any initial state $\mathbf{x}(t_0)$ and any bounded input $u(t)$, continuous on $[0, \infty)$, the solution $\mathbf{x}(t)$ exists for all $t \geq 0$ and satisfies

$$\| \mathbf{x}(t) \| = \beta(\| \mathbf{x}(t_0) \|, t - t_0) + \gamma \left(\sup_{t \leq \tau \leq t_0} \| u(\tau) \| \right)$$

for all t_0 and t such that $0 \leq t_0 \leq t$.

Definition 2.6 (Control Lyapunov Function (CLF) [11]). Consider the nonlinear system

$$\dot{\mathbf{x}} = f(\mathbf{x}) + g(\mathbf{x})\mathbf{u}, \tag{2.7}$$

and the time derivative of a Lyapunov function

$$\dot{V} = \frac{\partial V}{\partial \mathbf{x}} f(\mathbf{x}) + \frac{\partial V}{\partial \mathbf{x}} g(\mathbf{x}) = L_f V + L_g V \mathbf{u}.$$

A smooth, positive definite radially unbounded function $V(\mathbf{x})$ is called a

Control Lyapunov Function (CLF) if

$$\inf_{u \in \mathbb{R}^m} \{L_f V + L_g V \mathbf{u}\} < 0, \quad \forall \mathbf{x} \neq 0.$$

Lemma 2.2. *A smooth positive definite radially unbounded function V is a CLF if and only if* $L_g V = 0 \Rightarrow L_f V < 0, \forall \mathbf{x} \neq 0.$

Definition 2.7 (Input-to-State Stable-Control Lyapunov Function (ISS-CLF) [11]). A smooth positive definite radially unbounded function $V : \mathbb{R}^n \to \mathbb{R}_+$ is called an Input-to-State Stable-Control Lyapunov Function (ISS-CLF) for (2.29) if there exists a class-$\mathcal{K}_\infty$ function ρ such that the following implication holds for all $\mathbf{x} \neq 0$ and all $d \in \mathbb{R}^r$:

$$\|\mathbf{x}\| \geq \rho(\|d\|) \Rightarrow \inf_{u \in \mathbb{R}^m} \{L_f V + L_{g_1} V d + L_{g_2} V u\} < 0$$

Definition 2.8. A function $f(\mathbf{x}, t)$ is Lipschitz continuous in $\mathbf{x}$ with Lipschitz constant $L_C^f > 0$ if

$$\|f(\mathbf{x}, t) - f(\mathbf{y}, t)\| \geq L_C^f \|\mathbf{x} - \mathbf{y}\|$$

for all $\mathbf{x}, \mathbf{y}, t$.

Definition 2.9 (Lyapunov exponents). The Lyapunov exponents of $\dot{\mathbf{x}}(t) = f(\mathbf{x}, t)$ with initial condition x_0 at t_0 are defined by

$$h_i = \lim_{t \to \infty} \frac{1}{t} \ln \|\sigma_i(t)_i\|, \qquad i = 1, 2, \ldots, n,$$

when the limit exists, where $\sigma_i(t)_i$ are the eigenvalues of

$$Df(\mathbf{x}) = \left[\frac{\partial f}{\partial \mathbf{x}}\right]_{\mathbf{x}=0}.$$

Lyapunov exponents of a dynamical system measure the average local contraction or expansion in the phase space. Numerical algorithms for approximating Lyapunov exponents can be found in [17].

2.2   Chaotic Systems

The concept of chaos has recently attracted a great deal of attention and is one of the most rapidly expanding research topics in recent decades. Chaos is often associated with strange attractors or fractals. A chaotic system must have at least one positive Lyapunov exponent. According to the Poincaré–Bendixson Theorem, a continuous-time autonomous system must have dimension three or higher to be able to display chaos [22]. Chaotic attractors are used in this book to illustrate the effectiveness of control schemes due to the richness and complexity of the dynamics of these systems.

Some representative examples of chaotic systems are described in the following.

2.2.1   Lorenz system

The Lorenz system corresponds to the simplified equations derived from a mathematical model for atmospheric convection, where x, y, and z denote variables proportional to convective intensity, horizontal, and vertical temperature differences [15]. This system is given by

$$
\begin{aligned}
\dot{x} &= a_L(y - x), \\
\dot{y} &= x(b_L - z) - y, \\
\dot{z} &= xy - c_L z,
\end{aligned}
\tag{2.8}
$$

where a_L, b_L, and c_L are the Prandtl number, Rayleigh number, and a geometric factor, respectively. When $a_L = 10$, $b_L = 28$, and $c_L = 8/3$, the system presents a chaotic behaviors, which are used for numerical simulations. The corresponding attractor is displayed in Fig. 2.1(a).

2.2.2 Chen system

The Chen oscillator is described by [4]

$$
\begin{aligned}
\dot{x} &= a_C(y - x), \\
\dot{y} &= (c_C - a_C)x - xz + c_C y, \\
\dot{z} &= xy - b_C z,
\end{aligned}
\tag{2.9}
$$

with $a_C = 35$, $b_C = 3$, and $c_C = 28$, the system presents a chaotic behavior, and the corresponding attractor is displayed in Fig. 2.1(b).

This system is nonequivalent to the Lorenz system. It is dual to the Lorenz system in the sense of Čelikovskỳ and Vaněček [3]. It has even more sophisticated dynamical behaviors than the Lorenz system, as can be realized by comparing their attractors.

2.2.3 Lü system

The Lü system [14] is defined as

$$
\begin{aligned}
\dot{x} &= a_{Lu}(y - x), \\
\dot{y} &= -xz + c_{Lu}y, \\
\dot{z} &= xy - b_{Lu}z,
\end{aligned}
\tag{2.10}
$$

with $a_{Lu} = 36$, $b_{Lu} = 3$ and $c_{Lu} = 20$, which generates chaos, as shown in Fig. 2.1(c).

The Lü system represents the transition in the duality between the Lorenz and the Chen systems.

2.2.4   Chua's circuit

Chua's circuit is a third-order autonomous system, which can be easily realized in electronic form and exhibits a wide variety of nonlinear and chaotic phenomena [16]. The state equations are given by

$$
\begin{aligned}
\dot{x} &= P_C * (G_C \, (y - x - f_C(x))) \\
\dot{y} &= G_C(x - y) + z \\
\dot{z} &= -Q_C y,
\end{aligned}
\tag{2.11}
$$

with

$$
f_C(x) = b_C x + \frac{1}{2}(a_C - b_C)(|x + 1| - |x - 1|).
$$

Chua's circuit generates a double-scroll chaotic attractor with $P_C = 9$, $G_C = 0.7$, $Q_C = 7$, $a_C = -0.8$, and $b_C = -0.5$, as observed in Fig. 2.1(d).

2.2.5   Rössler system

The Rössler chaotic system [18] is given by

$$
\begin{aligned}
\dot{x} &= -(y + z), \\
\dot{y} &= x + a_R y, \\
\dot{z} &= b_R + z(x - c_R).
\end{aligned}
\tag{2.12}
$$

When $a_R = 0.2$, $b_R = 0.2$ and $c_R = 5.7$, the system is chaotic, and its attractor is displayed in Fig. 2.1(e).

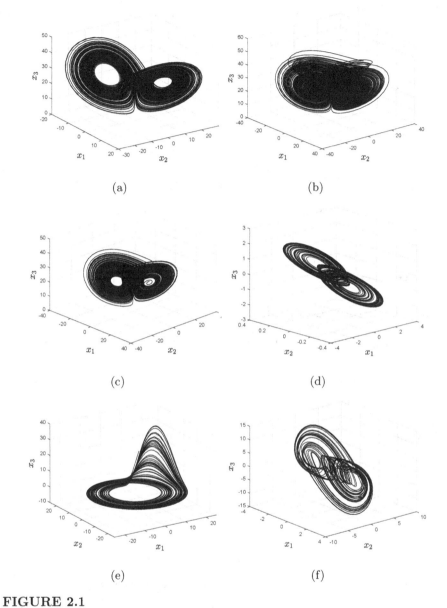

FIGURE 2.1

Chaotic attractor of (a) Lorenz system; (b) Chen system; (c) Lü system; (d) Chua's circuit; (e) Rössler system; (f) Arneodo system.

2.2.6   Arneodo system

The Arneodo system [1] is described by

$$\dot{x} = y,$$
$$\dot{y} = z,$$
$$\dot{z} = -a_A x - b_A y - c_A z + d_A x^3.$$

When $a_A = -5.5$, $b_A = 3.5$, $c_A = 1$ and $d_A = -1$, the system has a chaotic attractor as presented in Fig. 2.1(f).

2.3   Complex Dynamical Networks

The following definitions and lemmas are taken from [13, 5, 6].

Definition 2.10. A function $v : \mathbb{R}^n \to \mathbb{R}^n$ is increasing if $(\mathbf{x} - \mathbf{y})^T (v(\mathbf{x}, t) - v(\mathbf{y}, t)) \geq 0$ for all $\mathbf{x}, \mathbf{y}, t$.

Definition 2.11. A function $v : \mathbb{R}^n \to \mathbb{R}^n$ is uniformly increasing if there exists $c > 0$ such that

$$(\mathbf{x} - \mathbf{y})^T (v(\mathbf{x}, t) - v(\mathbf{y}, t)) \geq c\|\mathbf{x} - \mathbf{y}\|^2$$

for all $\mathbf{x}, \mathbf{y}, t$.

Definition 2.12. Given a square matrix V, a function $v : \mathbb{R}^n \to \mathbb{R}^n$ is V-uniformly increasing if Vv is uniformly increasing.

Definition 2.13. A function $v : \mathbb{R}^n \to \mathbb{R}^n$ is (V-uniformly) decreasing if $-v$ is (V-uniformly) increasing.

Definition 2.14. A function $v : \mathbb{R}^n \to \mathbb{R}^n$ is uniformly increasing if there exists $c > 0$ such that

$$(\mathbf{x} - \mathbf{y})^T (\upsilon(\mathbf{x}, t) - \upsilon(\mathbf{y}, t)) \geq c \|\mathbf{x} - \mathbf{y}\|^2,$$

for all $\mathbf{x}, \mathbf{y}, t$.

Lemma 2.3. *Suppose $f(\mathbf{x}, t)$ is Lipschitz continuous in $\mathbf{x}$ with Lipschitz constant $L_C^f > 0$ and V is symmetric and positive definite matrix of adequate dimensions. If*

$$\alpha > \frac{L_C^f}{\sigma_{min}(V)}$$

where $\sigma_{min}(V)$ is the smallest eigenvalue of V, then $\dot{\mathbf{x}} = f(\mathbf{x}, t) - \alpha V \mathbf{x}$ is asymptotically stable.

Proof. Consider $V = \frac{1}{2} x^T x$. The derivative of V is

$$
\begin{aligned}
\dot{V} &= x^T \left(f(x, t) - \alpha V x \right) \\
&= x^T f(x, t) - \alpha x^T V x \\
&\leq L_C^f \|x\|^2 - \alpha \sigma_{min}(V) \|x\|^2, \quad \|f(x, t)\| < L_C^f \|x\| \\
&\leq -\left(\alpha \sigma_{min}(V) - L_C^f \right) \|x\|^2 \\
&\leq -\rho \|x\|^2 < 0, \quad \rho > 0, \quad \alpha > \frac{L_C^f}{\sigma_{min}(V)}.
\end{aligned}
$$

$\square$

Definition 2.15 (Kronecker product). The Kronecker product of matrices A and B is defined as

$$A \otimes B = \begin{bmatrix} a_{11}B & \cdots & a_{1m}B \\ \vdots & \ddots & \vdots \\ a_{n1}B & \cdots & a_{nm}B \end{bmatrix}$$

where if $A \in \mathbb{R}^{n \times m}$ matrix and $B \in \mathbb{R}^{p \times q}$ matrix, then $A \otimes B \in \mathbb{R}^{np \times mq}$ matrix.

Definition 2.16. Vector $A \otimes f(\mathbf{x}_i, t)$ is defined as

$$A \otimes f(\mathbf{x}_i, t) = \begin{bmatrix} a_{11}f(\mathbf{x}_1, t) + a_{12}f(\mathbf{x}_2, t) + \cdots + a_{1m}f(\mathbf{x}_m, t) \\ \vdots \\ a_{n1}f(\mathbf{x}_1, t) + a_{n2}f(\mathbf{x}_2, t) + \cdots + a_{nm}f(\mathbf{x}_m, t) \end{bmatrix}$$

where if $A \in \mathbb{R}^{n \times m}$ matrix and $f \in \mathbb{R}^p$ is a function, then $A \otimes f(\mathbf{x}_i, t) \in \mathbb{R}^{np}$ is a vector.

Consider a weighted and undirected network with N linearly and diffusively coupled identical nodes, where each node is an n-dimensional dynamical system. The state equations of this dynamical network are given by

$$\dot{\mathbf{x}}_i = f(\mathbf{x}_i) + \sum_{j=1,\, j \neq i}^{N} c_{ij} a_{ij} \mathbf{\Gamma}(\mathbf{x}_j - \mathbf{x}_i), \quad i = 1, 2, \ldots, N, \tag{2.13}$$

where $\mathbf{x}_i = [x_{i_1}, x_{i_2}, \ldots, x_{i_n}]^T \in \mathbb{R}^n$ are the state variables of node i, $f : \mathbb{R}^n \to \mathbb{R}^n$ represents the self-dynamics of node i, the coupling strength between node i and node j is represented by constants $c_{ij} > 0$, the inner coupling matrix $\mathbf{\Gamma} = diag\{\tau_1, \tau_2, \ldots, \tau_n\}$ describes the connections among components of coupled nodes, and the coupling matrix $A = [a_{ij}] \in \mathbb{R}^{N \times N}$ stands for the network coupling configuration. If there is a connection between node i and node j for $i \neq j$, then $a_{ij} = a_{ji} = 1$; otherwise, $a_{ij} = a_{ji} = 0$ for $i \neq j$, and the diagonal elements of A are $a_{ii} = -k_i$ for $i = 1, 2, \ldots, N$, where k_i denotes the degree of node i, satisfying

$$\sum_{j=1, j \neq i}^{N} a_{ij} = \sum_{j=1, j \neq i}^{N} a_{ji} = k_i, \quad i = 1, 2, \ldots, N.$$

2.3.1 Pinning control strategy

For system (2.13), suppose that the control objective is to stabilize the whole network at a homogeneous stationary state $\bar{\mathbf{x}}$, which satisfies $f(\bar{\mathbf{x}}) = 0$:

$$\mathbf{x}_1 = \mathbf{x}_2 = \cdots = \mathbf{x}_N = \bar{\mathbf{x}}.$$

This objective is achieved by applying local linear feedback injections to a small fraction of the nodes in the network, which are called pinned. Assume that a fraction of nodes are pinned, which are labeled as $1, 2, \ldots, l$, where $1 \leq l \leq N$, and l can be as small as one [13]. Thus, the controlled network can be written as

$$
\begin{aligned}
\dot{\mathbf{x}}_i &= f(\mathbf{x}_i) + \sum_{j=1}^{N} c_{ij} a_{ij} \mathbf{\Gamma} \mathbf{x}_j + \mathbf{u}_i, & i &= 1, 2, \ldots, l, \\
\dot{\mathbf{x}}_i &= f(\mathbf{x}_i) + \sum_{j=1}^{N} c_{ij} a_{ij} \mathbf{\Gamma} \mathbf{x}_j, & i &= l+1, \ldots, N,
\end{aligned}
\tag{2.14}
$$

with

$$\mathbf{u}_i = -c_{ii} d_i \mathbf{T}(\mathbf{x}_i - \bar{\mathbf{x}}), \tag{2.15}$$

where $\mathbf{u}_i \in \mathbb{R}^n$ is the local feedback control law, $d_{k_i} > 0$ is the feedback gain, with the coupling strengths satisfying

$$c_{ii} a_{ii} + \sum_{j=1, j \neq i}^{N} c_{ij} a_{ij} = 0.$$

Define

$$
\begin{aligned}
D &= diag\,(d_{k_1}, d_{k_2}, \ldots, d_{k_l}, 0, \cdots, 0) \in \mathbb{R}^{N \times N}, \\
D' &= diag\,(c_{11} d_{k_1}, c_{22} d_{k_2}, \ldots, c_{ll} d_{k_l}, 0, \cdots, 0) \in \mathbb{R}^{N \times N}, \\
G &= (-c_{ij} a_{ij}) \in \mathbb{R}^{N \times N}.
\end{aligned}
$$

Moreover, using the Kronecker notation, write

$$\dot{\mathbf{X}} = I_N \otimes f(\mathbf{x}_i) - [(G + D) \otimes \mathbf{\Gamma}]\, \mathbf{X} + (D' \otimes \mathbf{\Gamma})\, \overline{\mathbf{X}}, \qquad (2.16)$$

where $\mathbf{X} = [\mathbf{x}_1^T, \mathbf{x}_2^T, \ldots, \mathbf{x}_N^T]^T \in \mathbb{R}^{Nn}$, $\overline{\mathbf{X}} = [\bar{\mathbf{x}}^T, \ldots, \bar{\mathbf{x}}^T]^T \in \mathbb{R}^{Nn}$, G is positive semi-definite and $G + D'$ is positive definite with the minimal eigenvalue $\sigma_{min}(G + D') > 0$.

Theorem 2.7 ([13]). *Suppose that $f(\mathbf{x})$ is Lipschitz continuous in $\mathbf{x}$ with Lipschitz constant $L_C^f > 0$ and $\mathbf{\Gamma}$ is symmetric and positive definite. If*

$$\sigma_{min}(G + D') > \alpha \equiv \frac{L_C^f}{\sigma_{min}(\mathbf{\Gamma})} > 0, \qquad (2.17)$$

where $\sigma_{min}(\mathbf{\Gamma})$ and $\sigma_{min}(G + D')$ are the minimal eigenvalue of matrices $\mathbf{\Gamma}$ and $G + D'$, respectively, then the equilibrium $\bar{\mathbf{x}}$ of the pinning controlled network (2.14) is globally asymptotically stable.

Condition (2.17) establishes a sufficient condition that the coupling strength matrix $C = [c_{ij}] \in \mathbb{R}^{N \times N}$ needs to satisfy in order to guarantee the equilibrium of network (2.14) to be globally and asymptotically stable. In the special case where $c_{ij} = c$ and $d_{k_i} = cd$, and $\mathbf{\Gamma} = I_n$, (2.17) reduces to

$$c > \frac{L_C^f}{\sigma_{min}(-A + diag(d_k, \ldots, d_k, 0, \ldots, 0))}. \qquad (2.18)$$

For a pinning controlled chaotic network (2.14), the Lyapunov exponent of each isolated node $\dot{\mathbf{x}}_i = f(\mathbf{x}_i)$ can characterize the local stability of the network, as follows.

Lemma 2.4 ([13]). *Assume that the node $\dot{\mathbf{x}}_i = f(\mathbf{x}_i)$ is chaotic for all $i = 1, 2, \ldots, N$, with the maximum positive Lyapunov exponent $h_{max} > 0$. If $c_{ij} = c$, $d_{k_i} = cd_k$, and $\mathbf{\Gamma} = I_n$, then the controlled network (2.14) is locally* *asymptotically stable about the homogeneous state* $\bar{\mathbf{x}}$, *provided that*

$$c > \frac{h_{max}}{\sigma_{min}\left(-A + diag(d_k, ..., d_k, 0, ..., 0)\right)}, \tag{2.19}$$

where σ_{min} *is the minimal eigenvalue of the matrix.*

2.3.2 Dynamical networks with non-identical nodes

Consider a controlled network consisting of N non-identical nodes with linear diffusive couplings given by

$$\dot{\mathbf{x}}_i = f_i(\mathbf{x}_i) + \sum_{j=1}^{N} c_{ij} a_{ij} \boldsymbol{\Gamma} \mathbf{x}_j + \mathbf{u}_i, \quad i = 1, 2, \ldots, N, \tag{2.20}$$

where $f_i : \mathbb{R}^n \to \mathbb{R}^n$ represents the self-dynamics of node i, and the other notations are defined as above. In order to pin this network to its equilibrium $\bar{\mathbf{x}}$, consider the following pinning controlled model

$$\begin{aligned}
\dot{\mathbf{x}}_i &= f_i(\mathbf{x}_i) + \sum_{j=1}^{N} c_{ij} a_{ij} \boldsymbol{\Gamma} \mathbf{x}_j + \mathbf{u}_i, \quad i = 1, 2, \ldots, l, \\
\dot{\mathbf{x}}_i &= f_i(\mathbf{x}_i) + \sum_{j=1}^{N} c_{ij} a_{ij} \boldsymbol{\Gamma} \mathbf{x}_j, \quad\quad i = l+1, \ldots, N.
\end{aligned} \tag{2.21}$$

The following assumption is needed to study network synchronization.

Assumption 1. *There is a continuously differentiable Lyapunov function* $V(x) : D \subseteq \mathbb{R}^n \mapsto \mathbb{R}_+$ *satisfying* $V(\bar{x}) = 0$ *with* $\bar{x} \in D$, *such that for each node function* $f_i(\mathbf{x}_i)$, *there is a scalar* θ_i *guaranteeing*

$$\frac{\partial V(\mathbf{x}_i)}{\partial \mathbf{x}_i}\left(f_i(\mathbf{x}_i) + g_i(\mathbf{x}_i, \bar{x}) + (\theta_i + \psi_i)\boldsymbol{\Gamma}\mathbf{x}_i\right) < 0,$$

$$\forall \mathbf{x}_i \in D_i, \quad \mathbf{x}_i \neq \bar{x}, \tag{2.22}$$

with constants $\psi_i \geq 0$, where

$$D_i = \{\mathbf{x}_i : \|\mathbf{x}_i - \bar{x}\| < \delta\}, \quad \delta > 0, \quad D = \bigcup_{i=1}^{N} D_i.$$

In (2.22), θ_i is the passivity degree [24]. This passivity degree is modified by a factor of ψ_i. Define the Lyapunov function for the controlled network:

$$V_N(\mathbf{X}) = \sum_{i=1}^{N} \frac{1}{2}\mathbf{x}_i^T \mathbf{P} \mathbf{x}_i, \tag{2.23}$$

$$\mathbf{X} = [\mathbf{x}_1^T, \mathbf{x}_2^T, \ldots, \mathbf{x}_N^T]^T.$$

By taking its time derivative, one obtains

$$\dot{V}_N(\mathbf{X}) = \sum_{i=1}^{N} \mathbf{x}_i^T \mathbf{P} \left(f_i(\mathbf{x}_i)) + \sum_{j=1}^{N} c_{ij}a_{ij}\mathbf{\Gamma}\mathbf{x}_j + g_i(\mathbf{x}_i, \bar{x}) \right)$$

$$< \sum_{i=1}^{N} \mathbf{x}_i^T \mathbf{P} \left(\sum_{j=1}^{N} c_{ij}a_{ij}\mathbf{\Gamma}\mathbf{x}_j - (\theta_i + \psi_i)\mathbf{\Gamma}\mathbf{x}_i \right)$$

$$< \mathbf{X}^T (-\mathbf{\Theta} + \mathbf{G} - \mathbf{\Psi}) \otimes \mathbf{P}\mathbf{\Gamma}\mathbf{X}. \tag{2.24}$$

The next result is established.

Theorem 2.8 ([24]). *If the coupling strength is sufficiently strong and ψ_i, $i = 1, 2, \ldots, l$, is large enough, then*

$$\mathbf{C} = -\mathbf{\Theta} + \mathbf{G} - \mathbf{\Psi}$$

is negative definite with nonzero l, where $\mathbf{\Theta} = diag(\theta_1, \theta_2, \ldots, \theta_N) \in \mathbb{R}^{N \times N}$, $\mathbf{G} = (g_{ij}) = (c_{ij}a_{ij}) \in \mathbb{R}^{N \times N}$, $\mathbf{\Psi} \in \mathbb{R}^{N \times N}$ is a diagonal matrix with the first l elements ψ_i, $i = 1, 2, \ldots, l$, and other $(N - l)$ elements are all zero.

The above method converts the original stability problem to the study of the negativity property of the matrix $\mathbf{C}$. This analysis method is named as V-stability [24].

2.4   Sliding-Mode Control

Following [21], consider the nonlinear system

$$\dot{\mathbf{x}} = f(\mathbf{x}) + g(\mathbf{x})u + \omega(\mathbf{x}), \tag{2.25}$$

where $\mathbf{x} \in \mathbb{R}^n$ is the state, $u \in \mathbb{R}^m$ is the control input, $\omega(\mathbf{x})$ characterizes the unknown disturbances; $f : \mathbb{R}^n \to \mathbb{R}^n$, $g : \mathbb{R}^n \to \mathbb{R}^{n \times m}$ are smooth nonlinear functions of $\mathbf{x}$, and $rank(g(\mathbf{x})) = m$ for all $\mathbf{x}$.

Two stages are important for the design of sliding-mode control. The first stage consists in selecting the sliding manifold $\mathbf{e}$, obtained as the intersection of m smooth manifolds:

$$\mathbf{e} = \{\mathbf{x} \in \mathbb{R}^n \mid e_i(\mathbf{x}) = 0, \quad i = 1, 2, \dots, m\},$$

where each manifold represents a desired system dynamics, which is of lower order than the given system. The second stage is to determine a discontinuous control law

$$u(\mathbf{x}) = \begin{cases} u^+(\mathbf{x}) & e_i(\mathbf{x}) > 0, \quad i = 1, 2, \dots, m \\ u^-(\mathbf{x}) & e_i(\mathbf{x}) < 0, \end{cases}$$

such that it drives the trajectories to the sliding manifold $(e_i(x) = 0)$ in finite time and then maintains them on this surface, such that the reaching modes satisfy the reachability condition.

One typical and classical approach for describing the dynamics of the sliding-mode is the *equivalent control method*, which is a smooth feedback control law, denoted by $u_{eq}(\mathbf{x})$. For this method, the conditions in the sliding-mode are analyzed at $e_i \equiv 0$ and $\dot{e}_i = 0$. To maintain the trajectory of the system on the manifold **e**, one obtains

$$\dot{e}_i(\mathbf{x}) = \frac{\partial e_i}{\partial \mathbf{x}^T}\left(f(\mathbf{x}) + g(\mathbf{x})u_{eq}(\mathbf{x} + \omega(\mathbf{x}))\right) = 0$$

$$L_f e_i(\mathbf{x}) + L_\omega e_i(\mathbf{x}) + L_g e_i(\mathbf{x})\,u_{eq}(\mathbf{x})\big|_{e_i=0} = 0.$$

Assuming that $L_{g_2}\mathbf{e} \neq 0$, for all **x**, the equivalent control is given by

$$u_{eq}(\mathbf{x}) = -\left[L_{g_2}e_i\right]^{-1}\left(L_f e_i\right)\mathbf{u}_{eq}(\mathbf{x} + L_\omega e_i(\mathbf{x}))\big|_{e_i=0}. \qquad (2.26)$$

Substituting (2.26) into (2.25) obtains the sliding-mode equation, given by

$$\dot{\mathbf{x}} = \left[I - g(\mathbf{x})\left[L_g e_i\right]^{-1}\frac{\partial e_i}{\partial \mathbf{x}^T}\right]f(\mathbf{x}) + \left[I - g(\mathbf{x})\left[L_g e_i\right]^{-1}\frac{\partial e_i}{\partial \mathbf{x}^T}\right]\omega(\mathbf{x})). \quad (2.27)$$

It follows from (2.27) that the ideal sliding dynamics is invariant with respect to disturbances, if $\omega(\mathbf{x}, t)$ satisfies the *matching condition*, $\omega(\mathbf{x}) \in span\, g(\mathbf{x})$, i.e.,

$$\omega(\mathbf{x}) = g(\mathbf{x})\vartheta(\mathbf{x}).$$

Such perturbations are called *matched perturbations*. In this case, (2.27) reduces to

$$\dot{\mathbf{x}} = \left[I - g(\mathbf{x})\left[L_g e_i\right]^{-1}\frac{\partial e_i}{\partial \mathbf{x}^T}\right]f(\mathbf{x}).$$

A classical control strategy, which achieves the control objective, i.e., to drive the trajectories to the sliding manifold ($e_i(x) = 0$) in finite time, is

$$u = \bar{u}_{eq} - K_c sign\,(e_i)\,, \qquad (2.28)$$

where $\bar{u}_{eq}$ represents the equivalent control (2.26) (without $L_\omega e_i$, since the disturbance is unknown), K_{ic} is a control gain, and

$$sign(e_i) = \begin{cases} 1, & e_i \geq 0 \\ -1, & e_i < 0. \end{cases}$$

2.5 Optimal Control

2.5.1 Continuous-time case

Consider the nonlinear system

$$\dot{\mathbf{x}} = f(\mathbf{x}) + g_1(\mathbf{x})\omega + g_2(\mathbf{x})u, \tag{2.29}$$

where $\mathbf{x} \in \mathbb{R}^n$ is the state, $\omega \in \mathbb{R}^r$ is the external input (e.g. disturbance), $u \in \mathbb{R}^m$ is the control input; $f : \mathbb{R}^n \to \mathbb{R}^n$, $g_1 : \mathbb{R}^n \to \mathbb{R}^{n \times r}$ and $g_2 : \mathbb{R}^n \to \mathbb{R}^{n \times m}$ are nonlinear functions of $\mathbf{x}$. The optimal robust stabilization control problem for system (2.29) is to find an optimal control law $u(t) = u^*$ that minimizes the cost functional

$$J(u) = \sup_{\omega \in \Omega} \left\{ \lim_{t \to \infty} \left[E(x(t)) + \int_0^t \left(l(\mathbf{x}) + u^T R_2(\mathbf{x}) u - \gamma(|\omega|) \right) d\tau \right] \right\}, \tag{2.30}$$

where Ω is the set of locally bounded functions of $\mathbf{x}$, which is solvable if there exist a class $\mathcal{K}_\infty$ function γ whose derivative γ' is also a class $\mathcal{K}_\infty$ function, $R_2 = R_2^T > 0$ is a matrix-valued function for all $\mathbf{x}$, $l(\mathbf{x})$ and $E(\mathbf{x})$ are radially unbounded functions.

For the direct approach of optimal control, the HJI equation must be solved. This can be avoided by using the inverse optimal control, introduced by Kalman [9] for linear systems. In this approach, a control law is first designed and then shown to be optimal for a cost functional of the form (2.30). The problem is inverse because the functions $l(\mathbf{x})$, $R_2(\mathbf{x})$ and $\gamma(|\omega|)$ are *a posteriori*

determined by the control law, rather than *a priori* chosen by the designer
[11].

Theorem 2.9 ([11]). *Consider the following auxiliary system of* (2.29):

$$\dot{\mathbf{x}} = f(\mathbf{x}) + g_1(\mathbf{x})\ell\gamma\left(2\left|L_{g_1}V\right|\right)\frac{(L_{g_1}V(\mathbf{x}))^T}{\left|L_{g_1}V\right|^2} + g_2(\mathbf{x})u, \qquad (2.31)$$

where $V(\mathbf{x})$ is a Lyapunov function and γ is a class $\mathcal{K}_\infty$ function whose derivate γ' is also a class $\mathcal{K}_\infty$ function. Suppose that there exists a matrix-valued function $R_2(\mathbf{x}) = R_2(\mathbf{x})^T > 0$ such that the control law $u = \alpha(\mathbf{x}) = -R_2(\mathbf{x})^{-1}(L_{g_2}V)^T$ globally asymptotically stabilizes (2.31) with respect to $V(\mathbf{x})$. Then, the control law

$$u = \alpha^*(\mathbf{x}) = \beta\alpha(\mathbf{x}) = -\beta R_2^{-1}(L_{g_2}V)^T \qquad (2.32)$$

with any $\beta \geq 2$ solves the inverse optimal gain assignment problem for system (2.29) *by minimizing the cost functional*

$$J(u) = \sup_{w\in\Omega}\left\{\lim_{t\to\infty}\left[2\beta V\left(\mathbf{x}(t)\right) + \int_0^t\left(l(\mathbf{x}) + u^T R_2\left(\mathbf{x}\right)u - \beta\lambda\gamma\left(\frac{|\omega|}{\lambda}\right)\right)d\tau\right]\right\} \qquad (2.33)$$

for any $\lambda \in (0, 2]$, where

$$l(\mathbf{x}) = -2\beta\left[L_fV + \ell\gamma\left(2\left|L_{g_1}V\right|\right) - L_{g_2}VR_2^{-1}(L_{g_2}V)^T\right]$$
$$+ \beta\left(2 - \lambda\right)\ell\gamma\left(2\left|L_{g_1}V\right|\right) + \beta\left(\beta - 2\right)L_{g_2}VR_2^{-1}(L_{g_2}V)^T. \qquad (2.34)$$

In Theorem 2.9, a sufficient condition for the solvability of the inverse optimal robust stabilization control problem is established. The existence of system (2.31) is ensured if system (2.29) is ISS, according to Definition 2.7, as shown in Section III of [11, pp. 338–342].

Remark 1. The control law (2.32), $u = \alpha_s(x)$, stabilizes the auxiliary system

(2.31), where

$$\alpha_s = \begin{cases} -\dfrac{\varpi + \sqrt{\varpi^2 + \left(L_{g_2}V\left(L_{g_2}V\right)^T\right)^2}}{L_{g_2}V\left(L_{g_2}V\right)^T}\left(L_{g_2}V\right)^T, & \left(L_{g_2}V\right)^T \neq 0 \\ 0, & \left(L_{g_2}V\right)^T = 0, \end{cases} \quad (2.35)$$

with

$$\varpi = L_f V + |L_{g_1}V|\rho^{-1}\left(|x|\right). \quad (2.36)$$

2.5.2 Discrete-time case

The next definition establishes the discrete-time inverse optimal control.

Definition 2.17 ([20]). The control law

$$u_k^* = -\frac{1}{2}R^{-1}(x_k)g^T(x_k)\frac{\partial V(x_{k+1})}{\partial x_{k+1}} \quad (2.37)$$

is inverse optimal (globally) stabilizing if:

- It achieves (globally) stability of $x = 0$ for the system

$$x_{k+1} = f(x_k) + g(x_k)u_k; \quad (2.38)$$

- $V(x_k)$ is a (radially unbounded) positive definite function satisfying inequality

$$\bar{V} := V(x_{k+1}) - V(x_k) + u_k^{*T}R(x_k)u_k^* \leq 0.$$

Note: Since the IOC problem is based on $V(x_k)$, one can choose

$$V(x_k) = \frac{1}{2}x_k^T P x_k, \quad (2.39)$$

where P is a matrix such that $P > 0$ and $P = P^T$. As a result, (2.37) becomes

$$u_k^* = \alpha(x_k) = -\frac{1}{2}\left[R(x_k) + \frac{1}{2}g^T(x_k)Pg(x_k)\right]^{-1}g^T(x_k)Pf(x_k). \quad (2.40)$$

Theorem 2.10. *Consider system* (2.38). *If there exists* $P = P^T > 0$ *such that*

$$V_f(x_k) - \frac{1}{4}f^T(x_k)Pg(x_k)[R(x_k) + \frac{1}{2}g^T(x_k)Pg(x_k)]^{-1}g^T(x_k)Pf(x_k) \leq \zeta_Q\|x_k\|^2,$$

where $V_f(x_k) = \frac{1}{2}[f^T(x_k)Pf(x_k) - V(x_k)]$ *and* $\zeta_Q > 0$ *denotes the minimum eigenvalue of matrix* Q, $\lambda_{\min}(Q)$, *then the equilibrium point* $x = 0$ *is globally exponentially stabilized by the control law* (2.40), *with the CLF* (2.39).

A meaningful cost function, which is minimized by (2.40), *is given by*

$$V(x_k) = \sum_{k=0}^{\infty}[\ell(x_k) + u_k^T R(x_k)u_k],$$

where $\ell(x_k) = -\bar{V}|_{u_k^* = \alpha(x_k)}$.

2.6 Recurrent High-Order Neural Networks

The use of multilayer neural networks is well-known for pattern recognition and for modeling of nonlinear functions. The NN is trained to learn an input-output map. Theoretical works have proven that, even with just one hidden layer, an NN can uniformly approximate any continuous function over a compact domain, provided that the NN has a sufficient number of synaptic connections.

For control tasks, extensions of the first-order Hopfield model called Recurrent High-Order Neural Networks (RHONN), which present more interactions among the neurons, are proposed in [19]. Additionally, the RHONN model is very flexible and allows us to incorporate in the neural model *a priori* information about the system structure.

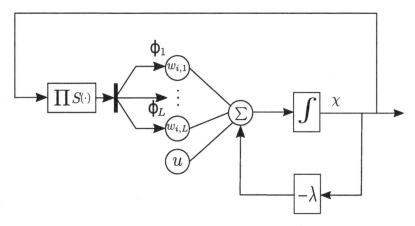

FIGURE 2.2

RHONN scheme.

2.6.1 Continuous-time model

An RHONN consists of n dynamical elements in the form of neurons and m external inputs, with high-order interactions among them. The state of each neuron is given by [19]

$$\dot{\chi}_i = -\lambda_i \chi_i + \sum_{k=1}^{L} w_{ik} \prod_{j \in I_k} y_j^{d_j(k)}, \qquad i = 1, 2, \ldots, n, \qquad (2.41)$$

where χ_i is the ith neuron state, L is the number of high-order connections, $\{I_1, I_2, \ldots, I_L\}$ is a collection of L non-ordered subsets of $\{1, 2, \ldots, m+n\}$, $\lambda_i > 0$, w_{ik} are the adjustable weights of the network, $d_j(k)$ are non-negative integers ($k = 1, 2, \ldots, L$, and $j \in I_k$), and y is the input vector to each neuron defined by

$$
\begin{aligned}
y &= [y_1, \ldots, y_n, y_{n+1}, \ldots, y_{n+m}]^T \\
&= [S(\chi_1), \ldots, S(\chi_n), S(u_1), \ldots, S(u_m)]^T, \qquad (2.42)
\end{aligned}
$$

with $u = [u_1, u_2, \ldots, u_m]$ being the input to the network, and $S(\cdot)$ being a smooth sigmoid function given by $S(\chi) = \frac{1}{1+e^{-\beta\chi}} + \epsilon$. For the sigmoid function, β is a positive constant and ϵ is a small positive constant, hence $S(\chi) \in [\epsilon, \epsilon + 1]$. As can be seen, (2.42) includes higher-order terms.

Now, define an L-dimensional vector $\phi(\cdot)$ by

$$\phi(\chi, u) = \left[\prod_{j \in I_1} y_j^{d_j(1)}, \prod_{j \in I_2} y_j^{d_j(2)}, \ldots, \prod_{j \in I_L} y_j^{d_j(L)} \right]^T . \tag{2.43}$$

Then, (2.41) can be written as

$$\dot{\chi}_i = -\lambda_i \chi_i + w_i \phi_i(\chi, u), \qquad i = 1, 2, \ldots, n, \tag{2.44}$$

where $w_i = [w_{i,1}, w_{i,2}, \ldots, w_{i,L}]^T$.

The RHONN used is affine in the control input, as displayed in Fig. 2.2. This RHONN in a matrix form is given as

$$\dot{\chi} = -\lambda I_n \chi + W\phi(\chi) + \mathbf{u}, \tag{2.45}$$

where $\chi \in \mathbb{R}^n$, $W \in \mathbb{R}^{n \times L}$, $\phi(\chi) \in \mathbb{R}^L$, $\mathbf{u} \in \mathbb{R}^n$, and $\lambda > 0$. Depending on the sigmoid function input, the RHONN can be classified as a series–parallel structure if $\phi(\cdot) = \phi(u)$, with u being an external input, or a parallel structure if $\phi(\cdot) = \phi(\chi)$, where χ is the neural network state [19]. These terminologies are standard in identification and adaptive control. For the application of the RHONN as an identifier, the series–parallel structure is used.

2.6.2   Discrete-time model

Consider the following discrete-time RHONN:

$$\chi_{i,k+1} = w_i^\top \phi_i(\chi_k, u_k), \quad i = 1, \ldots, n, \tag{2.46}$$

where χ_i $(i = 1, 2, \ldots, n)$ is the state of the ith neuron, L_i is the respective number of high-order connections, $\{I_1, I_2, \ldots, I_{L_i}\}$ is a collection of non-ordered subsets of $\{1, 2, \ldots, n + m\}$, n is the state dimension, m is the number of external inputs, w_i $(i = 1, 2, \ldots, n)$ is the respective on-line adapted weight vector, and $\phi_i(x_k, u_k)$ is given by

$$\phi_i(x_k, u_k) = \begin{bmatrix} \phi_{i_1} \\ \phi_{i_2} \\ \vdots \\ \phi_{i_{L_i}} \end{bmatrix} = \begin{bmatrix} \Pi_{j \in I_1} y_{i_j}^{d_{i_j}(1)} \\ \Pi_{j \in I_2} y_{i_j}^{d_{i_j}(2)} \\ \vdots \\ \Pi_{j \in I_{L_i}} y_{i_j}^{d_{i_j}(L_i)} \end{bmatrix}, \tag{2.47}$$

with $d_{j_i,k}$ being non-negative integers, and y_i defined as follows:

$$y_i = \begin{bmatrix} y_{i_1} \\ \vdots \\ y_{i_1} \\ y_{i_{n+1}} \\ \vdots \\ y_{i_{n+m}} \end{bmatrix} = \begin{bmatrix} S(x_1) \\ \vdots \\ S(x_n) \\ S(u_1) \\ \vdots \\ S(u_m) \end{bmatrix}. \tag{2.48}$$

In (2.48), $u = [u_1, u_2, \ldots, u_m]^\top$ is the input vector to the neural network, and

$$S(\varsigma) = \frac{1}{1 + \exp(-\beta\varsigma)}, \quad \beta > 0, \tag{2.49}$$

where ς is any real value variable.

Consider the problem to approximating the general discrete-time nonlinear system, by the following RHONN series–parallel representation:

$$\chi_{i,k+1} = w_i^{*\top} \phi_i(x_k, u_k) + \omega_{z_i}, \quad i = 1, \ldots, n, \tag{2.50}$$

where x_i is the ith plant state, ω_{z_i} is a bounded approximation error. Assume that there exists an ideal weight vector w_i^* such that $\|\omega_{z_i}\|$ can be minimized on a compact set $\Omega_{z_i} \subset \mathbb{R}^{L_i}$. The ideal weight vector w_i^* is an artificial quantity required for analytical purposes. In general, it is assumed that this vector exists and is constant but unknown. Next, define its estimate as w_i and the estimation error as

$$\widetilde{w}_{i,k} = w_i^* - w_{i,k} \,. \tag{2.51}$$

The estimate w_i is used for stability analysis, which will be discussed later. Since w_i^* is constant, one has $\widetilde{w}_{i,k+1} - \widetilde{w}_{i,k} = w_{i,k+1} - w_{i,k}$, $\forall k \in 0 \cup \mathbb{Z}^+$.

2.6.3   EKF training algorithm

The most well-known training approach for recurrent neural networks (RNNs) is backpropagation through time learning [23]. However, it is a first-order gradient descent method and hence its learning speed could be very slow [12]. Recently, Extended Kalman Filter (EKF)-based algorithms have been introduced to train neural networks [7]. With an EKF-based algorithm, the learning convergence is improved [12]. The EKF training of neural networks, both feedforward and recurrent ones, has proven to be reliable and practical for many applications over the past ten years [7].

It is known that Kalman filtering (KF) estimates the state of a linear system with an additive state and output white noise [8]. For EKF-based neural network training, the network weights become the states to be estimated. In this case, the error between the neural network output and the measured plant output can be considered as additive white noise.

The training goal is to determine the optimal weight values which minimize the prediction error. The EKF-based training algorithm is described by [8]:

$$
\begin{aligned}
K_{i,k} &= P_{i,k} H_{i,k} \left[R_{i,k} + H_{i,k}^\top P_{i,k} H_{i,k} \right]^{-1} \\
w_{i,k+1} &= w_{i,k} + \xi_i K_{i,k} \left[y_k - \hat{y}_k \right] \\
P_{i,k+1} &= P_{i,k} - K_{i,k} H_{i,k}^\top P_{i,k} + Q_{i,k},
\end{aligned}
\tag{2.52}
$$

where $P_i \in \mathbb{R}^{L_i \times L_i}$ is the prediction error associated with the covariance matrix, $w_i \in \mathbb{R}^{L_i}$ is the weight (state) vector, L_i is the total number of neural network weights, $y \in \mathbb{R}^m$ is the measured output vector, $\widehat{y} \in \mathbb{R}^m$ is the network output, ξ_i is a design parameter, $K_i \in \mathbb{R}^{L_i \times m}$ is the Kalman gain matrix, $Q_i \in \mathbb{R}^{L_i \times L_i}$ is the state noise associated covariance matrix, $R_i \in \mathbb{R}^{m \times m}$ is the measurement noise associated covariance matrix, and $H_i \in \mathbb{R}^{L_i \times m}$ is a matrix for which each entry (H_{ij}) is the derivative of one of the neural network outputs, $(\widehat{y})$, with respect to one neural network weight, (w_{ij}), as follows:

$$H_{ij,k} = \left[\frac{\partial \widehat{y}_k}{\partial w_{ij,k}} \right]_{w_{i,k} = \widehat{w}_{i,k+1}} \quad , \quad i = 1, ..., n \text{ and } j = 1, ..., L_i. \tag{2.53}$$

Usually P_i, Q_i and R_i are initialized as diagonal matrices, with entries $P_i(0)$, $Q_i(0)$ and $R_i(0)$, respectively. It is important to note that $H_{i,k}$, $K_{i,k}$ and $P_{i,k}$ for the EKF are bounded. Therefore, there exist constants $\overline{H_i} > 0$, $\overline{K_i} > 0$ and $\overline{P_i} > 0$ such that:

$$\begin{aligned}
\|H_{i,k}\| &\leq \overline{H_i} \\
\|K_{i,k}\| &\leq \overline{K_i} \\
\|P_{i,k}\| &\leq \overline{P_i}.
\end{aligned} \tag{2.54}$$

Remark 2. The measurement and process noises are typically characterized as zero-mean white noises with covariances given by $\delta_{k,j} R_{i,k}$ and $\delta_{k,j} Q_{i,k}$, respectively, with $\delta_{k,j}$ a Kronecker delta function (zero for $k \neq l$ and 1 for $k = l$). In order to simplify the notation in this book, the covariances will be represented by their respective associated matrices, $R_{i,k}$ and $Q_{i,k}$, for the noises and $P_{i,k}$ for the prediction errors.

Bibliography

[1] A. Arneodo, P. Coullet, E. Spiegel, and C. Tresser. Asymptotic chaos. *Physica D: Nonlinear Phenomena*, 14(3):327–347, 1985.

[2] S. P. Bhat and D. S. Bernstein. Finite-time stability of continuous autonomous systems. *SIAM Journal on Control and Optimization*, 38(3):751–766, 2000.

[3] S. Čelikovský and A. Vaněček. Bilinear systems and chaos. *Kybernetika*, 30(4):403–424, 1994.

[4] G. Chen and T. Ueta. Yet another chaotic attractor. *International Journal of Bifurcation and Chaos*, 9(7):1465–1466, 1999.

[5] G. Chen, X. Wang, and X. Li. *Fundamentals of Complex Networks*. Singapore: John Wiley & Sons, 2014.

[6] G. Chen. Pinning control and controllability of complex dynamical networks. *International Journal of Automation and Computing*, 14(1):1–9, 2017.

[7] L. Feldkamp, D. Prokhorov, and T. Feldkamp. Simple and conditioned adaptive behavior from Kalman filter trained recurrent networks. *Neural Networks*, 16:683–689, 2003.

[8] R. Grover and P. Hwang. *Introduction to Random Signals and Applied Kalman Filtering*, 2nd ed. New York, NY, USA: John Wiley and Sons, 1992.

[9] R. E. Kalman. When is a linear control system optimal? *Journal of Basic Engineering*, 86(1):51–60, 1964.

[10] H. K. Khalil. *Nonlinear Systems,* 2nd ed. Upper Saddle River, NJ, USA: Prentice Hall, 1996.

[11] M. Krstic and Z. H. Li. Inverse optimal design of input-to-state stabilizing nonlinear controllers. *IEEE Transactions on Automatic Control*, 43(3):336–350, 1998.

[12] C. Leunga and L. Chan. Dual extended Kalman filtering in recurrent neural networks. *Neural Networks*, 16:223–239, 2003.

[13] X. Li, X. Wang, and G. Chen. Pinning a complex dynamical network to its equilibrium. *IEEE Transactions on Circuits and Systems I: Regular Papers*, 51(10):2074–2087, 2004.

[14] J. Lü and G. Chen. A new chaotic attractor coined. *International Journal of Bifurcation and Chaos*, 12(3):659–661, 2002.

[15] E. Lorenz. Deterministic nonperiodic flow. *Journal of the Atmospheric Sciences*, 20(2):130–141, 1963.

[16] T. Matsumoto, L. Chua, and M. Komuro. The double scroll. *IEEE Transactions on Circuits and Systems*, 32(8):797–818, 1985.

[17] T. Parker and L. Chua. *Practical Numerical Algorithms for Chaotic Systems.* New York, NY, USA: Springer-Verlag, 1989.

[18] O. E. Rössler. An equation for continuous chaos. *Physics Letters A*, 57(5):397–398, 1976.

[19] G. A. Rovithakis and M. A. Christodoulou. *Adaptive Control with Recurrent High-Order Neural Networks.* London, UK: Springer-Verlag, 2000.

[20] E. N. Sanchez and F. Ornelas-Tellez. *Discrete-Time Inverse Optimal Control for Nonlinear Systems.* Boca Raton, FL, USA: CRC Press, 2017.

[21] V. Utkin, J. Guldner, and M. Shijun. *Sliding Modes in Control and Optimization.* Berlin, Germany: Springer-Verlag, 2013.

[22] S. Wiggins. *Introduction to Applied Nonlinear Dynamical Systems and Chaos.* New York, NY, USA: Springer-Verlag, 1990.

[23] R. J. Williams and D. Zipser. A learning algorithm for continually running fully recurrent neural networks. *Neural Computation*, 1:270–280, 1989.

[24] J. Xiang and G. Chen. On the V-stability of complex dynamical networks. *Automatica*, 43(6):1049–1057, 2007.

Part II

Sliding-Mode Control

3

Model-Based Sliding-Mode Control

In this chapter, a novel approach named sliding-mode pinning control is introduced to achieve trajectory tracking of complex networks. Two cases are presented. For the first case the whole network tracks a reference for each one of the states; the second case uses the backstepping technique to track a desired trajectory for only one state. The illustrative example is composed of a network of 50 nodes; each node dynamics is a Chen chaotic attractor.

3.1   Sliding-Mode Pinning Control

3.1.1   Case 1

The following scheme visualized in Fig. 3.1; it is composed by a reference system for the network and a controller for each pinned node.

To establish a control law, based on the sliding-mode technique, consider a complex network with pining control as in (2.14). Define $\mathbf{x}_s$ as the desired nonlinear reference system given by

$$\dot{\mathbf{x}}_s = f_s(\mathbf{x}_s), \quad \mathbf{x}_s \in \mathbb{R}^n. \tag{3.1}$$

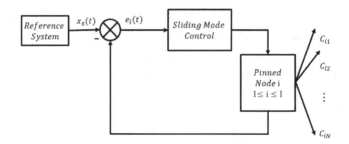

FIGURE 3.1
The proposed control scheme.

For trajectory tracking, suppose that the pinned node dynamics is known. Then, the tracking error is defined as $\mathbf{e}_i = \mathbf{x}_i - \mathbf{x}_s$.

The following theorem is established:

Theorem 3.1. *Consider the complex network with pinning control as in (2.14), with $\mathbf{\Gamma} = \mathbf{I}_n$, a reference system defined by (3.1) with $f_s(\mathbf{x}_s)$ being a chaotic function, h_{max} being the maximum positive Lyapunov exponent of $f_i(\mathbf{x}_i) - f_s(\mathbf{x}_s)$. If c_{min}, the minimal coupling strength of the whole network, fulfills*

$$c_{min} > \frac{h_{max}}{\sigma_{min}(-A_0 + \hat{D}_0)}, \qquad (3.2)$$

where A_0 is obtained from the coupling matrix $A = [a_{ij}] \in \mathbb{R}^{N \times N}$ by removing those rows and columns that correspond to the controlled nodes $i = 1, 2, \ldots, l$, and $\hat{D}_0 = diag(\hat{d}_{0_{l+1}}, \ldots, \hat{d}_{0_N})$ with $\hat{d}_{0_i} = \sum\limits_{j=1}^{l} a_{ij}$, then the sliding-mode pinning control law given by

$$\mathbf{u}_i = \bar{\mathbf{u}}_{eq_i} - K_{s_i}\, sign\,(\mathbf{e}_i(\mathbf{x}_i)), \quad i = 1, 2, \ldots, l, \qquad (3.3)$$

with $\bar{\mathbf{u}}_{eq_i} = f_s(\mathbf{x}_s) - f(\mathbf{x}_i)$, $K_{s_i} \in \mathbb{R}$ a control gain, and $\mathbf{e}_i \in \mathbb{R}^n$ the desired sliding manifold, ensures that the tracking error along the trajectories of (3.1), is locally asymptotically stable.

Proof. Substitute the control law (3.3) into (2.14) to get

$$\dot{\mathbf{e}}_i = \sum_{j=1}^{N} c_{ij} a_{ij} \boldsymbol{\Gamma} \mathbf{x}_j - K_{s_i} \, sign\,(\mathbf{e}_i(\mathbf{x}_i)), \quad i = 1, 2, \ldots, l,$$

$$\triangleq f(\mathbf{x}) + g(\mathbf{x})\mathbf{u} + \omega(\mathbf{x}),$$

where $f(\mathbf{x}) = 0$, $g(\mathbf{x}) = I_n$, and $\omega(\mathbf{x}) = \sum_{j=1}^{N} c_{ij} a_{ij} \boldsymbol{\Gamma} \mathbf{x}_j$ are the influence of other network nodes, which are considered as a disturbance.

Assume now that the following bound is satisfied:

$$\left\| \sum_{j=1}^{N} c_{ij} a_{ij} \boldsymbol{\Gamma} \mathbf{x}_j \right\|_1 \leq q_i, \quad q_i > 0, \quad i = 1, 2, \ldots, l. \tag{3.4}$$

Select the controller (3.3), with

$$K_{s_i} > q_i + \frac{\Upsilon_i}{\sqrt{n}}, \quad \Upsilon_i > 0, \quad i = 1, 2, \ldots, l.$$

Defining the Lyapunov function

$$V(\mathbf{e}_i) = \frac{1}{2} \mathbf{e}_i^T \mathbf{e}_i \Rightarrow 2V = \mathbf{e}_i^T \mathbf{e}_i \Rightarrow \sqrt{2V} = \|\mathbf{e}_i\|_2,$$

and taking its time derivative, one gets

$$\dot{V}(\mathbf{e}_i) = \mathbf{e}_i^T \left(\sum_{j=1}^{N} c_{ij} a_{ij} \boldsymbol{\Gamma} \mathbf{x}_j - K_{s_i} \, sign\,(\mathbf{e}_i) \right)$$

$$\dot{V}(\mathbf{e}_i) \leq \left\| \dot{V}(\mathbf{e}_i) \right\|_1$$

$$\leq \|S_i\|_1 \left\| \sum_{j=1}^{N} c_{ij} a_{ij} \boldsymbol{\Gamma} \mathbf{x}_j \right\|_1 - K_{s_i} \|\mathbf{e}_i\|_1 \| \, sign\,(\mathbf{e}_i)\,\|_1$$

$$\leq \|\mathbf{e}_i\|_1 \left\| \sum_{j=1}^{N} c_{ij} a_{ij} \boldsymbol{\Gamma} \mathbf{x}_j \right\|_1 - K_{s_i} \|\mathbf{e}_i\|_1.$$

Using (3.4) and $\| \cdot \|_2 \leq \| \cdot \|_1 \leq \sqrt{n} \, \| \cdot \|_2$ [p. 648][3], one obtains

$$
\begin{aligned}
\dot{V}(\mathbf{e}_i) &\leq -(K_{s_i} - q_i)\sqrt{n} \, \|\mathbf{e}_i\|_2 \\
&\leq -\Upsilon_i \, \|\mathbf{e}_i\|_2 \\
&\leq -\Upsilon_i \sqrt{2} \, (V(\mathbf{e}_i))^{1/2} < 0.
\end{aligned}
\tag{3.5}
$$

Since $V(\mathbf{e}_i) > 0$ is positive definite, the real number $c = \Upsilon_i \sqrt{2} > 0$ for $K_{s_i} > q_i$, and $\alpha = \frac{1}{2}$, it follows that

$$
\dot{V}(\mathbf{e}_i) + \Upsilon_i \sqrt{2} \, (V(\mathbf{e}_i))^{1/2} \leq 0
$$

is satisfied.

According to Theorem 2.5, the control law (3.3) guarantees the convergence of the pinned node states to the manifold $\mathbf{e}_i \equiv 0$ in finite time defined by the settling time function

$$
T(\sigma_i) \leq \frac{2}{\Upsilon_i \sqrt{2}} V(\sigma_i)^{1/2} \, .
$$

In order to analyze the sliding modes dynamics stability ($\mathbf{e}_i = \mathbf{x}_i - \mathbf{x}_s \equiv 0$, $\dot{\mathbf{e}}_i = 0$), the sliding mode equation is written as

$$
\begin{aligned}
\mathbf{x}_i &= \mathbf{x}_s, & i &= 1, 2, \ldots, l, \\
\dot{\mathbf{x}}_i &= f(\mathbf{x}_i) + \sum_{j=l+1}^{N} c_{ij} a_{ij} \mathbf{\Gamma} \mathbf{x}_j + \sum_{j=1}^{l} c_{ij} a_{ij} \mathbf{\Gamma} \mathbf{x}_s, & i &= l+1, \ldots, N.
\end{aligned}
\tag{3.6}
$$

Considering

$$
a_{ii} = -\sum_{j=l, j \neq i}^{N} a_{ij} = -\left(\sum_{j=l+1, j \neq i}^{N} a_{ij} + \sum_{j=1}^{l} a_{ij} \right),
\tag{3.7}
$$

network (3.6) can be rewritten as

$$
\mathbf{x}_i = \mathbf{x}_s, \qquad\qquad\qquad i = 1, 2, \ldots, l,
$$

$$
\dot{\mathbf{x}}_i = f(\mathbf{x}_i) + \sum_{j=l+1}^{N} c_{ij} a_{0_{ij}} \mathbf{\Gamma} \mathbf{x}_j - \sum_{j=1}^{l} c_{ij} a_{ij} \mathbf{\Gamma}(\mathbf{x}_i - \mathbf{x}_s), \qquad i = l+1, \ldots, N,
$$

$$(3.8)$$

where $A_0 = [a_{0_{ij}}] \in \mathbb{R}^{(N-l)\times(N-l)}$ is obtained from the coupling matrix A by removing those rows and columns, which correspond to the controlled nodes $i = 1, 2, \ldots, l$, i.e.,

$$
a_{0_{ij}} = a_{ij}, \qquad\qquad j \neq i, \quad j = l+1, \ldots, l, \quad i = l+1, \ldots, l,
$$

$$
a_{0_{ij}} = - \sum_{j=l+1, j\neq i}^{N} a_{ij}, \quad i = l+1, \ldots, N.
$$

The proof follows the same procedure as in [7, 9]. The controlled network (3.8) is linearized for the unpinned nodes $i = l+1, \ldots, N$, so that

$$
\dot{\eta} = \eta \left[D f_{is} \left(\mathbf{x}_s \right) \right] - \hat{B}\eta,
$$

where $D f_{is} \left(\mathbf{x}_s \right) \in \mathbb{R}^{n\times n}$ is the Jacobian of $(f_i - f_s)$ at $\mathbf{x}_s$,

$$
\eta = [\eta_{l+1}, \eta_{l+2}, \ldots, \eta_N]^T \in \mathbb{R}^{(N-l)n},
$$

with $\eta_i(t) = \mathbf{x}_i(t) - \mathbf{x}_s(t)$, $i = l+1, l+2, \ldots, N$, and $\hat{B} = (G_0 + \hat{D}) \in \mathbb{R}^{(N-l)\times(N-l)}$, where $G_0 = [-c_{ij} a_{0_{ij}}] \in \mathbb{R}^{(N-l)\times(N-l)}$, $\hat{D} = diag(\hat{d}_{l+1}, \ldots, \hat{d}_N)$ with $\hat{d}_i = \sum_{j=1}^{l} a_{ij} c_{ij}$. According to [10], one has

$$
0 < \sigma_{\min}\left(c_{\min}\left[-A_0 + \hat{D}_0 \right] \right) \le \sigma_{\min}\left(G_0 + \hat{D} \right),
$$

where $\hat{D}_0 = diag(\hat{d}_{0_{l+1}}, \ldots, \hat{d}_{0_N})$ with $\hat{d}_{0_i} = \sum_{j=1}^{l} a_{ij}$. The above inequality is obtained based on the fact that $c_{ij} \ge c_{min} > 0$ for all c_{ij} in G_0. Here, the coupling strengths can be different for different nodes.

Furthermore, the Transversal Lyapunov Exponents (TLEs) denoted by $\mu_k(\sigma_i)$, for each eigenvalue σ_i, $i = l+1, l+2, \ldots, N$, is given by [6]

$$\mu_k(\sigma_i) = h_k - c_{ij}\sigma_i, \quad k = 1, 2, \ldots, n,$$

where h_k is the respective Lyapunov exponent. The TLEs determine the stability of the controlled states [9], hence the local stability of the controlled network (2.14), ensures negative TLEs. Thus, the following condition must be satisfied:

$$\mu_{\max}(\sigma_{\min}) = h_{\max} - c_{\min}\sigma_{min}(-A_0 + \hat{D}_0) < 0, \tag{3.9}$$

inequality (3.9) is equivalent to condition (3.2). Then, with the control law (3.3), trajectory tracking is achieved.

□

3.1.2   Case 2

Consider to control the network (2.14) to track an output reference $\mathbf{x}_s \in \mathbb{R}^m$ ($m < n$) with pinning control $\mathbf{u}_i \in \mathbb{R}^m$. Assume that the pinned node dynamics is known. The tracking error is defined as $\mathbf{e}_i = \mathbf{x_i} - \mathbf{x}_s$. In this case, the backstepping technique is used to design a sliding manifold such that the resulting sliding mode dynamics is described by a desired linear system. Next, one can synthesize a discontinuous control law, which enforces sliding-mode motion into the sliding manifold by using backstepping control. For more details about backstepping control see [4].

This approach is feasible if the nonlinear system can be transformed to a special state-space form, named as block feedback form, given by the following definition.

Definition 3.1 (Block-feedback form (BFF) [4]). The system

$$
\begin{aligned}
\dot{\mathbf{x}}_j &= f_j(\bar{\mathbf{x}}_j) + g_j(\bar{\mathbf{x}}_j)x_{j+1} + \omega_j(t), \quad j = 1, 2, \ldots, r - 1, \\
\dot{\mathbf{x}}_r &= f_r(\mathbf{x}) + g_r(\mathbf{x})u + \omega_r(t), \\
\dot{\mathbf{x}}_{r+1} &= f_{r+1}(\mathbf{x}) + g_{r+1}(\mathbf{x})u + \omega_{r+1}(t), \\
y &= \mathbf{x}_1,
\end{aligned}
\tag{3.10}
$$

where $\mathbf{x} = [\mathbf{x}_1, \mathbf{x}_2, \ldots, \mathbf{x}_{r+1}]^T$, $\mathbf{x}_{r+1}$ represents the zero dynamics of the system, $\bar{\mathbf{x}}_j = [\mathbf{x}_1, \mathbf{x}_2, \ldots, \mathbf{x}_j]^T$, $j = 1, 2, \ldots, r$, $\mathbf{x}_j \in \mathbb{R}^{nj}$, r is the number of blocks, $\omega(t) \in \mathbb{R}^{nj}$ is the bounded unknown disturbance vector, then there is a constant $\bar{\omega}_j$ such that $\|\omega_j(t)\| \leq \bar{\omega}_j$, for $0 < t < \infty$, system (3.10) has the block-feedback form with Zero dynamics, if

$$
rank\ [g_j] = n_j \quad \forall \mathbf{x} \in \mathbb{R}^n, \forall t \in [0, \infty), \quad j = 1, 2, \ldots, r,
$$

The numbers $n_1, n_2, \ldots, n_r$ are the controllability indexes satisfying

$$
n_1 \leq n_2 \leq \ldots \leq n_r \leq m,
$$

with $\sum_{j=1}^{r+1} n_j = n$.

The following theorem is established to guarantee output tracking for the whole network, using sliding-mode pinning control.

Theorem 3.2. *Assume that the complex network* (2.14), *with pinning control law defined as*

$$
\mathbf{u}_i = \bar{\mathbf{u}}_{eq_i} - K_{r_i} sign\left(z_{r_i}(\mathbf{x}_i)\right) \quad \in \mathbb{R}^m, \quad i = 1, 2, \ldots, l,
\tag{3.11}
$$

where $\bar{\mathbf{u}}_{eq_i}$ *is the equivalent control,* K_{r_i} *is a control gain, and* $z_{r_i} \in \mathbb{R}^m$ *is the desired sliding manifold, can be transformed to BFF. If its zero dynamics is stable and condition* (3.2) *is fulfilled, then the tracking error, along the trajectories of* $\mathbf{x}_s$, *is locally ultimately bounded.*

Proof. Consider system (2.14) transformed to the BFF form as

$$
\begin{aligned}
\dot{\mathbf{x}}_{i_j} &= f_{ij}(\overline{\mathbf{x}}_{i_j}) + g_{ij}(\overline{\mathbf{x}}_{i_j})\mathbf{x}_{j+1_i} + \omega_{ij}(t), && j = 1, 2, \ldots, r-1 \wedge i = 1, 2, \ldots, l, \\
\dot{\mathbf{x}}_{i_r} &= f_{ir}(\mathbf{x}_i) + g_{ir}(\mathbf{x}_i)u + \omega_{ir}(t), && i = 1, 2, \ldots, l, \\
\dot{\mathbf{x}}_{i_{r+1}} &= f_{ir+1}(\mathbf{x}_i) + g_{ir+1}(\mathbf{x}_i)u + \omega_{ir+1}(t), && i = 1, 2, \ldots, l, \\
\dot{\mathbf{x}}_i &= f(\mathbf{x}_i) + \sum_{j=1}^{N} c_{ij} a_{ij} \boldsymbol{T}\mathbf{x}_j, && i = l+1, \ldots, N,
\end{aligned}
$$

$$(3.12)$$

where $\omega_{ij}(t) = \sum_{j=1}^{N} c_{ij} a_{ij} \boldsymbol{\Gamma}\mathbf{x}_j$, and $\bar{\omega}_{ij} \geq |\omega_{ij}(t)|$, $\forall i = 1, 2, \ldots, N, \forall j = 1, 2, \ldots, n$, are considered disturbances. Initially, the sliding manifold is designed for each pinned node ($i = 1, 2, \ldots, l$) using the backstepping technique, which is described step-by-step as follows.

Step 1: Let z_1 be the error between $\mathbf{x}_1$ and its desired value $\mathbf{x}_s$:

$$z_1 = \mathbf{x}_1 - \mathbf{x}_s.$$

Define the Lyapunov function as

$$V(z_1) = \frac{1}{2}\|z_1\|^2 > 0,$$

$$\dot{V}(z_1) = z_1^T \dot{z}_1 = z_1^T (f_1(\bar{\mathbf{x}}_1) + g_1(\bar{\mathbf{x}}_1)x_2 + \omega_1(t) - \dot{\mathbf{x}}_s). \tag{3.13}$$

The objective is to design a virtual control $\mathbf{x}_2 = \alpha_1$, which forces $z_1 \to 0$. This control is

$$\alpha_1 = g_1^{-1}(\bar{\mathbf{x}}_1)(-f_1(\bar{\mathbf{x}}_1) + \dot{\mathbf{x}}_r - k_1 z_1) \qquad \text{with} \qquad k_1 > 0.$$

Then, (3.13) becomes

$$\dot{V}(z_1) = -k_1\|z_1\|^2 + z_1^T \omega_1 + g_1(\bar{\mathbf{x}}_1)z_1^T z_2.$$

Step j: The error dynamics for z_j is

$$z_j = \mathbf{x}_j - \alpha_{j-1},$$

which represents the error between the actual and the virtual control inputs. Select the augmented Lyapunov function as

$$V(z_1,\ldots,z_j) = \frac{1}{2}\|z_1\|^2 + \cdots + \frac{1}{2}\|z_j\|^2 = V(z_1,\ldots,z_{j-1}) + \frac{1}{2}\|z_j\|^2,$$

and its derivative

$$
\begin{aligned}
\dot{V}(z_1,\ldots,z_j) &= z_1^T \dot{z}_1 + \cdots + z_j^T \dot{z}_j \qquad\qquad (3.14)\\
&= \dot{V}(z_1,\ldots,z_{j-1}) + z_j^T\left(f_j(\bar{\mathbf{x}}_j) + g_j(\bar{\mathbf{x}}_j)x_{j+1} + \omega_j(t) - \dot{\alpha}_{j-1}\right).
\end{aligned}
$$

Analogously, the objective at the j-th step is to design a virtual control $x_{j+1} = \alpha_j$ in order to stabilize the error $z_j = 0$, which is

$$\alpha_j = g_j^{-1}(\bar{\mathbf{x}}_j)\left(-f_j(\bar{\mathbf{x}}_j) - \dot{\alpha}_{j-1} - g_{j-1}(\bar{\mathbf{x}}_{j-1})z_{j-1} - k_j z_j\right), \quad \text{with} \quad k_j > 0. $$

$$(3.15)$$

Replacing (3.15) in (3.14) gives

$$\dot{V}(z_1,\ldots,z_j) = -\sum_{k=1}^{j} k_k \|z_k\|^2 + \sum_{k=1}^{j} z_k^T \omega_k + g_j(\bar{\mathbf{x}}_j)z_j z_{j+1}.$$

Step r: The sliding manifold is selected as $z_r = \mathbf{x}_s - \alpha_{r-1}$. The dynamics for the pinned nodes (3.12) in z-variables is

$$\dot{z}_1 \;=\; -k_1 z_1 + g_1(\bar{\mathbf{x}}_1) z_2 + \omega_1,$$

$$\vdots$$

$$\dot{z}_j \;=\; -g_{j-1}(\bar{\mathbf{x}}_{j-1}) z_{j-1} - k_j z_j + g_j(\bar{\mathbf{x}}_j) z_{j+1} + \omega_j, \qquad j = 2, 3, \ldots, r-1,$$

$$\vdots$$

$$\dot{z}_{r-1} \;=\; -g_{r-2}(\bar{\mathbf{x}}_{r-2}) z_{r-2} - k_{r-1} z_{r-1} + g_{r-1}(\bar{\mathbf{x}}_{r-1}) z_r + \omega_{r-1},$$

$$\dot{z}_r \;=\; f_r(\mathbf{x}) + g_r(\mathbf{x}) u + \omega_r(t) - \dot{\alpha}_{r-1},$$

$$\dot{\mathbf{x}}_{r+1} \;=\; f_{r+1}(\mathbf{x}) + g_{r+1}(\mathbf{x}) u + \omega_{r+1}(t).$$

Now, defining u as a discontinuous feedback control

$$u \;=\; g_r^{-1}(\mathbf{x})\left(-f_r(\mathbf{x}) + \dot{\alpha}_{r-1}\right) - g_r^{-1}(\mathbf{x}) k_r \, sign(z_r)$$

$$u \;=\; \bar{u}_{eq} - K_r \, sign \, (z_r(\mathbf{x}_i)),$$

where $\bar{u}_{eq} = g_r^{-1}(\mathbf{x})\left(-f_r(\mathbf{x}) + \dot{\alpha}_{r-1} - g_{r-1}(\bar{\mathbf{x}}_{r-1}) z_{r-1}\right)$, and $K_r = g_r^{-1}(\mathbf{x}) k_r$, a Lyapunov function is selected as $V(z_r) = \frac{1}{2} z_r^T z_r$, and the derivative of $V(z_r)$ is

$$\dot{V}(z_r) \;=\; z_r^T \dot{z}_r$$

$$=\; z_r(-k_r \, sign \, (z_r) + \omega_r)$$

$$\leq\; -k_r \|z_r\|_1 + \bar{\omega}_r \|z_r\|_1 \quad (k_r \geq \frac{u_0}{\sqrt{n_r}} + \bar{\omega}_r)$$

$$\leq\; -u_0 \|z_r\|_2 \leq -u_0 \sqrt{2}(V(z_r))^{\frac{1}{2}} < 0.$$

Then, control law (3.11) guarantees the convergence to the manifold $z_r = 0$ in finite time.

To analyze the stability on sliding modes, the sliding-mode equation for the whole network is rewritten as

$$
\begin{aligned}
\dot{\boldsymbol{\zeta}}_i &= A_i \boldsymbol{\zeta}_i + E_i(\boldsymbol{\zeta}_i) + \boldsymbol{\omega}_i, & i &= 1, \ldots, l, \\
\dot{\mathbf{x}}_{i_{r+1}} &= f_{ir+1}(\mathbf{x}_i) + g_{ir+1}(\mathbf{x}_i)\bar{u}_{eq_i}, & & \\
\dot{\mathbf{x}}_i = f(\mathbf{x}_i) + \sum_{j=l+1}^{N} c_{ij} a_{0_{ij}} \boldsymbol{\Gamma} \mathbf{x}_j &- \sum_{j=1}^{l} c_{ij} a_{ij} \boldsymbol{\Gamma}(\mathbf{x}_i - \boldsymbol{\alpha}), & i &= l+1, \ldots, N,
\end{aligned}
$$

$$(3.16)$$

where $\boldsymbol{\zeta}_i = [z_i^{(1)}, \ldots, z_i^{(r-1)}]^T$, $A_i = diag(-k_i^{(1)}, \ldots, -k_i^{(r-1)})$, $\boldsymbol{\omega}_i = [\omega_i^{(1)}, \ldots, \omega_i^{(r-1)}]^T$,

$$
E_i(\boldsymbol{\zeta}_i) = [g_{i1}(\bar{\mathbf{x}}_{i_1})z_i^{(2)}, -g_{i1}(\bar{\mathbf{x}}_{i_1})z_i^{(1)} + g_{i2}(\bar{\mathbf{x}}_{i_2})z_i^{(3)}, \ldots, -g_{ir-2}(\bar{\mathbf{x}}_{i_{r-2}})z_i^{(r-2)}],
$$

correspond to the crossed terms, $\mathbf{x}_{i_{r+1}}$ correspond to zero dynamics of pinned nodes, $\boldsymbol{\alpha} = [\mathbf{x}_s, \alpha_1, \ldots, \alpha_{r-1}]^T$. $\boldsymbol{\zeta}_i$ can be consider as a linear stable system, perturbed by the last terms. Since these terms are bounded, is possible to conclude that the trajectories of $\boldsymbol{\zeta}$ are ultimately bounded. Moreover, it is assumed that the zero dynamics for pinned nodes is stable. For stability of unpinned nodes ($i = l+1, \ldots, N$), first, define $\mathbf{z}_i = \mathbf{x}_i - \boldsymbol{\alpha}$; the dynamics of these nodes is linearized on an outer neighborhood of the ultimate bound, so that

$$
\dot{\eta} = \eta \left[Df_i(\boldsymbol{\alpha}) \right] - \hat{B}\eta,
$$

where $Df_i(\boldsymbol{\alpha}) \in \mathbb{R}^{n \times n}$ is the Jacobian of f_i at $\boldsymbol{\alpha}$, and

$$
\eta = [\eta_{l+1}, \eta_{l+2}, \ldots, \eta_N]^T \in \mathbb{R}^{(N-l)n},
$$

with $\eta_i(t) = \mathbf{z}_i(t)$, $i = l+1, l+2, \ldots, N$.

Therefore, the same procedure as Theorem 3.1 can be applied in the linear region, which implies that the tracking error for the whole network is ultimately bounded. $\square$

3.2   Simulation Results

To illustrate tracking performance and dynamical behavior of the controlled network, two cases are presented. For the first one, the whole network tracks a reference for each one of the states by means of the control inputs for pinned nodes $(u_i(t) \in \mathbb{R}^n, \quad i = 1, 2, \ldots, l)$; on the other hand, the second case uses the backstepping technique to track a desired trajectory for only one state; in this case the control is scalar for each pinned nodes $(u_i(t) \in \mathbb{R}, \quad i = 1, 2, \ldots, l)$.

3.2.1   Case 1

Consider a scale-free network of chaotic Chen's oscillators [1] with degree distribution $\delta(K_i) \approx k_i^{-2}$ and $N = 50$ nodes. Fig. 3.2 presents phase portraits of Chen's oscillator for different parameters: 1. ($a_C = 35$, $b_C = 3$, and $c_C = 28$), 2. ($a_C = 30$, $b_C = 3$, and $c_C = 20$), and 2. ($a_C = 50$, $b_C = 4$, and $c_C = 40$). Suppose that $\mathbf{\Gamma} = diag(1,1,1)$ and the example network tracks the desired reference system (3.1) applying the sliding-mode pinning controller only to the node with the highest connection number ($i = 1$). Defining the state variables as: $x_1 = x$, $x_2 = y$, $x_3 = z$, the equation of the pinned node $\mathbf{x}_1$ is

$$\dot{x}_{1_1} = a_C(x_{1_2} - x_{1_1}) + \sum_{j=1}^{50} c_{1j}a_{1j}\mathbf{\Gamma}x_{j_1} + u_{1_1}$$

$$\dot{x}_{1_2} = (c_C - a_C)x_{1_1} - x_{1_1}x_{1_3} + c_C x_{1_2} + \sum_{j=1}^{50} c_{1j}a_{1j}\mathbf{\Gamma}x_{j_2} + u_{1_2} \quad (3.17)$$

$$\dot{x}_{1_3} = x_{1_1}x_{1_2} - b_C x_{1_3} + \sum_{j=1}^{50} c_{1j}a_{1j}\mathbf{\Gamma}x_{j_3} + u_{1_3},$$

where $\mathbf{u}_1 = f_s(\mathbf{x}_s) - f(\mathbf{x}_1) - K_{s_1}\, sign\,(\mathbf{e}_1(\mathbf{x}_1))$, with $\mathbf{e}_1(\mathbf{x}_1) = \mathbf{x}_1 - \mathbf{x}_s$.

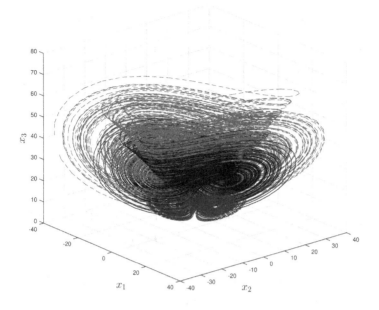

FIGURE 3.2
Phase portrait of Chen's oscillator for different parameters.

Simulations are performed using Matlab/Simulink with the Euler solver and a fixed step size of 0.5×10^{-4}. Simulations are performed as follows. From $t = 0s$ to $t = 4.8s$, the network runs without any connection ($c_{ij} = 0$). From $t = 4.8s$, the coupling strengths are a random set and is time-variant, with $c_{ij} > 20$. Then, at $t = 5s$, the control law is applied; the network is stabilized to a constant reference, which is selected as the unstable equilibrium point, $\mathbf{x}_s = [7.9373, 7.9373, 21]^T$. Thereafter, at $t = 10s$, a reference selected as a the chaotic Lorenz attractor [8] is incepted to generate the desired trajectory $\mathbf{x}_s(t)$. From $t = 15s$, plant parameters are changed for the network odd nodes as follows:

$$a_C = \begin{cases} 30, & 15 < t \le 16 \quad \text{and} \quad 19 < t \le 20 \\ 35, & 0 < t \le 15, \ 16 < t \le 17, \quad \text{and} \quad 18 < t \le 19 \\ 50, & 17 < t \le 18 \end{cases}$$

$$b_C = \begin{cases} 3, & 0 < t \le 17 \quad \text{and} \quad 18 < t \le 20 \\ 4, & 17 < t \le 18 \end{cases}$$

$$c_C = \begin{cases} 20, & 15 < t \le 16 \quad \text{and} \quad 19 < t \le 20 \\ 28, & 0 < t \le 15, \ 16 < t \le 17, \quad \text{and} \quad 18 < t \le 19 \\ 40, & 17 < t \le 18 \end{cases}$$

For all the described events, c_{ij} fulfills always equation (3.2). Fig. 3.3 displays the state evolutions of the entire network. Before $t = 5s$, trajectory evolves freely without control action; when the proposed control law is applied, the complex network tracks the desired trajectory. As can be seen, tracking is achieved for both a constant reference and a chaotic one even in the presence of plant parameters variations, with a mean square error (MSE) of 0.75, illustrating the robustness of the controller. Fig. 3.4 shows the control input signal $u_i(t)$ as applied to the pinned node, displaying the typical chattering characteristic for discontinuous control actions based on sliding modes. To eliminate or at least reduce chattering, the boundary layer, observer-based, regular form, and disturbance rejection techniques can be used as in [2, 5, 11]. Moreover, Fig. 3.5 displays simulation results for the average trajectory error $\bar{e}_{s_i}(t)$, $i = 1, 2, 3$. Fig. 3.6 presents the values of the coupling strengths $c_{ii} = \frac{1}{k_i} \sum_{j=1, j \ne i}^{N} c_{ij} a_{ij}$ with c_{ii}, $i = 1, 2, \dots, 50$. Finally, Fig. 3.7 presents the plant parameter changes associated with the values presented in Fig. 3.2. The effects of the interconnected nodes and the coupling strengths variations can be seen as unmodeled dynamics and external disturbances. Based on theoretical and simulation results, it is concluded that the main advantage of the proposed controller is to achieve trajectory tracking successfully even in the presence of plant parameter changes, unmodeled dynamics, and bounded external disturbances.

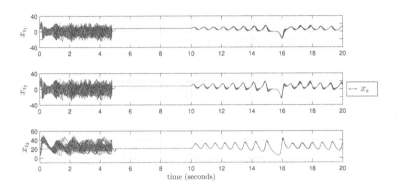

FIGURE 3.3
Network states evolutions for Case 1.

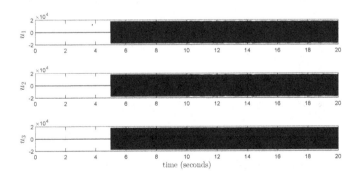

FIGURE 3.4
Control input signals for the pinned node.

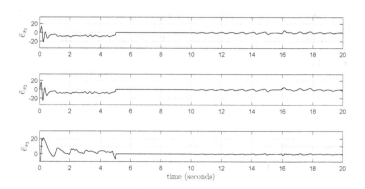

FIGURE 3.5
Average trajectory error.

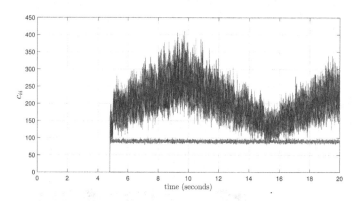

FIGURE 3.6
Coupling strengths.

3.2.2 Case 2

Consider a scalar control $u_i \in R$, which is applied to only one state of the whole network. The equation of the pinned node is

$$
\begin{aligned}
\dot{x}_{1_1} &= a_C(x_{1_2} - x_{1_1}) + \sum_{j=1}^{50} c_{1j} a_{1j} \mathbf{\Gamma} x_{j_1} \\
\dot{x}_{1_2} &= (c_C - a_C)x_{1_1} - x_{1_1} x_{1_3} + c_C x_{1_2} + \sum_{j=1}^{50} c_{1j} a_{1j} \mathbf{\Gamma} x_{j_2} \qquad (3.18) \\
\dot{x}_{1_3} &= x_{1_1} x_{1_2} - b_C x_{1_3} + \sum_{j=1}^{50} c_{1j} a_{1j} \mathbf{\Gamma} x_{j_3} + u_1,
\end{aligned}
$$

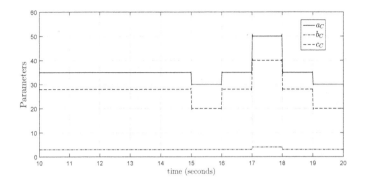

FIGURE 3.7
Time evolution of Chen's oscillator parameters (a_C, b_C, and c_C).

where $u_1 = \bar{u}_{eq} - k_{3_1} \, sign(z_{3_1})$, with

$$z_{1_1} = x_{1_1} - x_s$$

$$z_{1_2} = x_{1_2} - \alpha_1, \quad \alpha_1 = \frac{1}{a_C}(a_C x_{1_1} + x_{1_r} - k_1 z_{1_1})$$

$$z_{1_3} = x_{1_3} - \alpha_2, \quad \alpha_2 = \frac{1}{x_{1_1}}\left((c_C - a_C)x_{1_1} + c x_{1_2} - \dot{\alpha}_1 + a_C z_{1_1} + k_2 z_{1_2}\right),$$

and $\bar{u}_{eq} = -x_1 x_2 + b_C x_3 + \dot{\alpha}_2 + x_1 z_2$.

For simulations, x_{1_d}, the changes and other characteristics are the same as for Case 1. From $t = 0s$ to $t = 4.8s$, the network runs without any connection ($c_{ij} = 0$). From $t = 4.8s$, the coupling strengths are set constant, randomly chosen with $c_{ij} \geq 30$. Then, at $t = 5s$, the control law is applied; the state x_1 of the whole network is stabilized to a constant reference, which is selected as $\mathbf{x}_s = 7.9373$. Thereafter, at $t = 10s$, chaotic Lorenz system (2.8) ($\mathbf{x}_s = x_{s_1}(t)$) is selected as reference; furthermore, at $t = 15s$, the value of coupling strengths c_{ij} changes for 25% of nodes of the entire network.

All the described c_{ij} fulfill always equation (3.2). Fig. 3.8 displays the time response for the output $y = x_1$ (only one state of the whole network). As can be seen, tracking is achieved for both a constant reference and a chaotic one. Fig. 3.9 shows the control input signal u_3 applied to the pinned node; since

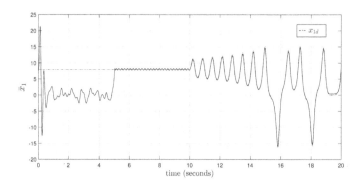

FIGURE 3.8
Time response for only one state of the whole network.

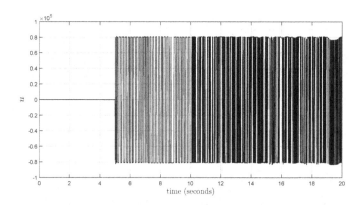

FIGURE 3.9
Control signal u_3 for the pinned node.

the whole network (50 nodes) tracks a desired trajectory controlling only one node, the required energy is large and depends on the number and location of the pinned nodes. Moreover, Fig. 3.10 shows the simulation results for the state evolutions. Before $t = 5s$, trajectory evolves freely without control; when the control law is applied, the state x_1 tracks the desired trajectory, and x_2 and x_3 track the trajectory given by Chen oscillator. Furthermore, at $t = 15s$, the disturbances are incepted as in Case 1, validating the robustness properties of the proposed controller with a MSE of 0.6921. The effects of

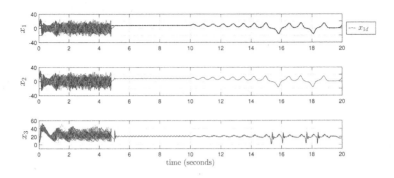

FIGURE 3.10
Network states evolutions for Case 2.

the interconnected nodes and the coupling strengths variations can be seen as unmodeled dynamics and external disturbances.

Differing from previously published results, this controller ensures output tracking using pinning control strategy with ultimately bounded tracking errors even in the presence of unmodeled dynamics and bounded external disturbances.

3.3 Conclusions

This chapter has presented a new control strategy for trajectory tracking on complex networks, based on pinning control and the sliding-mode technique. Simulation results illustrate that trajectory tracking for a scale-free network of chaotic nodes can be effectively achieved by using this control scheme. Two cases are discussed; the first case presents sliding-mode pinning control for the whole network to follow a reference for each one of the states, where this controller is applied on the pinned node to control 50 nodes (150 states). It is shown that trajectory tracking is achieved. The second case considers the control for only one state, applying backstepping technique; desired output tracking is also achieved.

Bibliography

[1] G. Chen and T. Ueta. Yet another chaotic attractor. *International Journal of Bifurcation and Chaos*, 9(7):1465–1466, 1999.

[2] J. Guldner and V. I. Utkin. The chattering problem in sliding mode systems. *14th Intenational Symposium of Mathematical Theory of Networks and Systems*, Perpignan, France, 11, 2000.

[3] H. Khalil. *Nonlinear Systems*. Upper Saddle River, NJ, USA: Prentice-Hall, 2002.

[4] M. Krstic, I. Kanellakopoulos, and P. Kokotovic. *Nonlinear and Adaptive Control Design*. Wiley, 1995.

[5] H. Lee and V. I. Utkin. Chattering suppression methods in sliding mode control systems. *Annual Reviews in Control*, 31(2):179–188, 2007.

[6] X. Li and G. Chen. Synchronization and desynchronization of complex dynamical networks: An engineering viewpoint. *IEEE Transactions on Circuits and Systems I: Fundamental Theory and Applications*, 50(11):1381–1390, 2003.

[7] X. Li, X. Wang, and G. Chen. Pinning a complex dynamical network to its equilibrium. *IEEE Transactions on Circuits and Systems I: Regular Papers*, 51(10):2074–2087, 2004.

[8] E. Lorenz. Deterministic nonperiodic flow. *Journal of the Atmospheric Sciences*, 20(2):130–141, 1963.

[9] H. Su and W. Xiaofan. *Pinning Control of Complex Networked Systems: Synchronization, Consensus and Flocking of Networked Systems via Pinning.* Berlin, Germany: Springer-Verlag, 2013.

[10] E. Sanchez and D. Rodriguez. Inverse optimal pinning control for complex networks of chaotic systems. *International Journal of Bifurcation and Chaos*, 25(02): 1550031, 2015.

[11] V. Utkin and H. Lee. Chattering problem in sliding mode control systems. *Variable Structure Systems. VSS'06. International Workshop on Variable Structure Systems*, 346–350, 2006.

4

Neural Sliding-Mode Control

In this chapter, a new approach to achieve output synchronization for uncertain complex networks with non-identical nodes using neural sliding-mode pinning control is introduced. The control scheme is composed by an on-line identifier based on a recurrent high-order neural network, and a sliding-mode controller, where the former is used to build an on-line model for the unknown dynamics, and the latter to force the unknown node dynamics to achieve output synchronization.

4.1 Formulation

The considered complex network model with N non-identical nodes and linear diffusive coupling, where each node is an n-dimensional dynamical system, is given by

$$
\begin{cases}
\dot{\mathbf{x}}_i & = \quad \hat{f}_i(\mathbf{x}_i) + \sum_{j=1}^{N} c_{ij} a_{ij} \mathbf{\Gamma} \mathbf{x}_j, \quad i = 1, 2, \ldots, N, \\
\mathbf{y}(t) & = \quad \mathcal{C} \mathbf{x}_i(t), \quad t \geq 0
\end{cases}
\tag{4.1}
$$

where $\mathbf{x}_i = [x_{i_1}, x_{i_2}, \ldots, x_{i_n}]^T \in \mathbb{R}^n$ is the state vector of the ith node i; $\hat{f}_i : \mathbb{R}^n \to \mathbb{R}^n$ represents the unknown self-dynamics of node i; $\mathbf{y} = [y_1, y_2, \ldots, y_q]^T \in \mathbb{R}^q$, refers to the output vector of node i, $\mathcal{C}$ is a known matrix with proper dimensions, constant c_{ij} is the coupling strength between node i and node j, $\boldsymbol{\Gamma} > 0 = diag(\gamma_1, \gamma_2, \ldots, \gamma_n) \in \mathbb{R}^{n \times n}$ is the inner coupling matrix, and $\mathbf{A} = [a_{ij}] \in \mathbb{R}^{N \times N}$ is the coupling configuration matrix representing the topological structure of network (4.1) whose elements are defined as follows: if there is a directed link from node j to node i $(i \neq j)$, then $a_{ij} = 1$, otherwise, $a_{ij} = 0$, and

$$a_{ii} = - \sum_{j=1, j \neq i}^{N} a_{ij}, \quad i = 1, 2, \ldots, N.$$

Thus, one has $\sum_{j=1}^{N} a_{ij} = 0$. Functions $f_i(\mathbf{x}_i)$ satisfy the next assumption.

Assumption 2. *[11] There is a continuously differentiable Lyapunov function* $V(\mathbf{z}_i) : D \subseteq \mathbb{R}^n \to \mathbb{R}_+$ *satisfying* $V(\mathbf{z}_i(0)) = 0$, *such that for each node function* $f_i(\cdot)$, *there is a scalar* ψ_i *guaranteeing*

$$\frac{\partial V(\mathbf{z}_i)}{\partial \mathbf{z}_i} \left(f_i(\mathbf{z}_i + \boldsymbol{\alpha}_i) - f_i(\boldsymbol{\alpha}_i) + \psi_i \boldsymbol{\Gamma} \mathbf{z}_i \right) < 0, \quad (4.2)$$

for all $\mathbf{z}_i = \mathbf{x}_i - \boldsymbol{\alpha}_i \in D_i$, $\mathbf{z}_i \neq 0$, *where*

$$D_i = \{\mathbf{z}_i : \|\mathbf{z}_i\| < \delta\}, \quad \delta > 0, \quad D = \bigcup_{i=1}^{N} D_i,$$

and ψ_i *represents the passivity degree.*

For this chapter, the matrix $\mathbf{A}$ is not assumed to be symmetric, which implies that the complex network can be a directed one. The control goal is to synthesize a controller to achieve output synchronization, defined as follows:

Definition 4.1. Network (4.1) is output synchronized if

$$\lim_{t \to \infty} \|\mathbf{y}(t) - \mathbf{y}_j(t)\| = 0,$$

for any $j = 1, 2, \ldots, N$.

To achieve this objective, the pinning control technique [6, 2] is used, which consists in applying local controllers to a small number of network nodes, referred to as pinned ones. Considering that the first l nodes are selected to be pinned, the controlled network is described by

$$
\text{pinned nodes} \begin{cases}
\dot{\mathbf{x}}_{i_1} &= \hat{f}_{i_1}(\mathbf{x}_i) - \sum_{j=1}^{N} c_{ij} a_{ij} \mathbf{\Gamma} \mathbf{x}_{j_1} \\
\dot{\mathbf{x}}_{i_2} &= \hat{f}_{i_2}(\mathbf{x}_i) - \sum_{j=1}^{N} c_{ij} a_{ij} \mathbf{\Gamma} \mathbf{x}_{j_2} + \mathbf{u}_i, \\
i &= 1, \ldots, l,
\end{cases}
$$

$$
\text{unpinned nodes} \begin{cases}
\dot{\mathbf{x}}_i &= \hat{f}_i(\mathbf{x}_i) - \sum_{j=1}^{N} c_{ij} a_{ij} \mathbf{\Gamma} \mathbf{x}_i, \\
i &= l+1, \ldots, N,
\end{cases}
$$

$$\mathbf{y} = \mathcal{C} \mathbf{x}_i, \tag{4.3}$$

where $\mathbf{x}_{i_1} \in \mathbb{R}^{n-m}$, $\mathbf{x}_{i_2} \in \mathbb{R}^m$, $\mathbf{u}_i \in \mathbb{R}^m$, vector functions $\hat{f}_{i_1}(\mathbf{x}_i)$ and $\hat{f}_{i_2}(\mathbf{x}_i)$ are unknown, in which $\mathbf{x}_i$ is available for measurement. The neural sliding-mode pinning control scheme is visualized in Fig. 4.1. In order to synthesize the control law, an on-line neural identifier based on an RHONN scheme is used; furthermore, the sliding-mode technique is applied to achieve output synchronization.

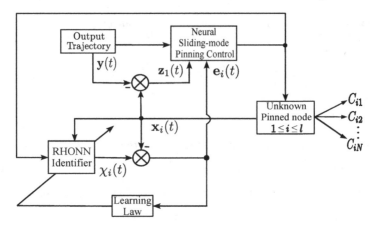

FIGURE 4.1
The neural control scheme.

4.2   Neural Identifier

Assume that the pinned nodes have a Block Controllable Form (BCF) . Then, an RHONN in a series-parallel structure [4, 8] consisting of r blocks for identifying the unknown nonlinear dynamics (4.3) is as follows:

$$\dot{\chi}_{i_j} = -\lambda \chi_{i_j} + W_{i_j} \phi_{i_j}(\mathbf{x}_{i_1}, \ldots, \mathbf{x}_{i_j}) + W'_{i_j} \mathbf{x}_{i_{j+1}},$$
$$\dot{\chi}_{i_r} = -\lambda \chi_{i_r} + W_{i_r} \phi_{i_r}(\mathbf{x}_i) + W'_{i_r} u_i,$$
$$j = 1, \ldots, r-1, \quad i = 1, \ldots, l, \tag{4.4}$$

where $\lambda > 0$, $\chi_i = [\chi_{i_1}, \chi_{i_2}, \ldots, \chi_{i_r}]^T \in \mathbb{R}^n$ is the ith neuron state vector, $W_{i_k} \in \mathbb{R}^{n_k \times L}$, $k = 1, \ldots, r$, is the on-line adjustable weight matrix, $W'_{i_k} \in \mathbb{R}^{n_{k+1} \times L}$, $k = 1, \ldots, r$, is the fixed weight matrix, $\phi_{i_k}(\cdot) \in \mathbb{R}^L$ defined before mentioned. The set of numbers $(n_1, \ldots, n_r)$ are the controllability indexes, which define the structure of system (4.4), and satisfy

$$n_1 \leq n_2 \leq \cdots \leq n_r \leq m,$$

with $\sum_{k=1}^{r} n_k = n$.

The following assumptions are needed [9]:

Assumption 3. *For every $w_{ik} \in W_i$, (4.4) is bounded for every bounded state* x_i.

Assumption 4. *There is an unknown optimal constant weight matrix W_i^* such that the pinned nodes (4.3) are described fully by an RHONN scheme, satisfying*

$$\dot{\mathbf{x}}_{i_j} = -\lambda \mathbf{x}_{i_j} + W_{i_j}^* \phi_{i_j}(\mathbf{x}_{i_1}, \ldots, \mathbf{x}_{i_j}) + W_{i_j}' \mathbf{x}_{i_{j+1}} + \boldsymbol{\omega}_{i_j},$$

$$\dot{\mathbf{x}}_{i_r} = -\lambda \mathbf{x}_{i_r} + W_{i_r}^* \phi_{i_r}(\mathbf{x}_i) + W_{i_r}' u_i + \boldsymbol{\omega}_{i_r},$$

$$\mathbf{y} = C\mathbf{x}_i,$$

$$j = 1, \ldots, r - 1, \quad i = 1, \ldots, l,$$

where the term $\boldsymbol{\omega}_i \in \mathbb{R}^n$ represents the modeling error, which is bounded, and all the elements are as described above.

A weight learning law is selected to minimize the identification error, including enough high-order connections. The identification error is given by $\varepsilon_i = \boldsymbol{\chi}_i - \mathbf{x}_i$, satisfying

$$\dot{\varepsilon}_i = -\lambda \varepsilon_{i_i} + \tilde{W}_i \phi(\mathbf{x_i}),$$

where $\tilde{W}_i = W_i - W_i^*$. The learning law is [8]

$$tr\left\{ \dot{\tilde{W}}_i^T \tilde{W}_i \right\} = -\gamma \varepsilon_i^T \tilde{W}_i \phi(\mathbf{x}_i),$$

with which, the identification error satisfies

$$\sup_{0 \leq t \leq T} \|\varepsilon_i(t)\| \leq \mathcal{E},$$

for any $\mathcal{E} > 0$ and any finite $T > 0$.

4.3 Output Synchronization

Suppose that the pinned nodes have been identified by the RHONN (4.4) after a finite time T [8]. However, this neural identifier cannot reduce the modeling error $\boldsymbol{\delta}_i$ to zero; hence, $\omega_i = \delta_i$ is considered as a disturbance, leading to

$$\dot{\mathbf{x}}_i \overset{\triangle}{=} \dot{\chi}_i + \omega_i, \tag{4.5}$$

$$\dot{\mathbf{x}}_{i_j} = -\lambda\chi_{i_j} + W_{i_j}\phi_{i_j}(\mathbf{x}_{i_1},\ldots,\mathbf{x}_{i_j}) + W'_{i_j}\mathbf{x}_{i_{j+1}} + \omega_{i_j},$$

$$\dot{\mathbf{x}}_{i_r} = -\lambda\chi_{i_r} + W_{i_r}\phi_{i_r}(\mathbf{x}_i) + W'_{i_r}u_i + \omega_{i_r},$$

$$j = 1,\ldots,r-1, \quad i = 1,\ldots,l.$$

The following theorem establishes conditions for achieving output synchronization.

Theorem 4.1. *Consider the complex network (4.3), with* $\boldsymbol{\Gamma} = I_n$, *and the sliding-mode pinning control law*

$$u_i = (W'_r)^{-1}\left(-f_r(\mathbf{x}_i, \chi_i, W_i)\right) - (W'_r)^{-1}k_r sign(\mathbf{z}_{r_i})$$

$$= \bar{u}_{eq} - K_r\, sign\,(\mathbf{z}_{r_i}), \tag{4.6}$$

where χ_i *are the identified states of the pinned nodes,*

$$\bar{u}_{eq} = (W'_r)^{-1}\left(-f_r(\mathbf{x}, \chi, W)\right),$$

$K_r = (W'_r)^{-1}k_r$, *and* $\mathbf{z}_{r_i} \in \mathbb{R}^m$ *is the desired sliding manifold. If the following condition is satisfied:*

$$M = (-\boldsymbol{\Psi} + \mathbf{G}) < 0, \tag{4.7}$$

where $\psi_i = (1 - \zeta)\sigma_{min}(A_i)$ *for the pinned nodes* $(i = 1, 2, \ldots, l)$, *and* ψ_i *is given by Assumption 2 for the unpinned nodes* $(i = l + 1, l + 2, \ldots, N)$, ζ *and* A_i *are selected parameters,* $\boldsymbol{\Psi} = diag(\psi_1, \psi_2, \ldots, \psi_N) \in \mathbb{R}^{N \times N}$, *and*

$\mathbf{G} = (g_{ij}) = (c_{ij}a_{ij}) \in \mathbb{R}^{N \times N}$, *then the output synchronization error along the desired trajectories of* $\mathbf{y}$ *is ultimately bounded.*

Proof. To achieve the control objective, the sliding manifold is defined, for each pinned node $(i = 1, 2, \ldots, l)$ using the backstepping technique [5], as follows.

Step 1: Define the tracking error $\mathbf{z}_1$ by

$$\mathbf{z}_1 = \mathbf{x}_1 - \mathbf{y}.$$

Select the Lyapunov function as

$$V(\mathbf{z}_1, \varepsilon_1, \tilde{W}_1) = \frac{1}{2} \mathbf{z}_1^T \mathbf{z}_1 + \frac{1}{2} \varepsilon_1^T \varepsilon_1 + \frac{1}{2\gamma} tr\left\{\tilde{W}_1^T \tilde{W}_1\right\} > 0,$$

which gives

$$\begin{aligned}
\dot{V}(\mathbf{z}_1, \varepsilon_1, \tilde{W}_1) =& \mathbf{z}_1^T \dot{\mathbf{z}}_1 + \varepsilon_1^T \dot{\varepsilon}_1 + \frac{1}{\gamma} tr\left\{\dot{\tilde{W}}_1^T \tilde{W}_1\right\} \\
=& \mathbf{z}_1^T (-\lambda \boldsymbol{\chi}_1 + W_1 \phi_1(\mathbf{x}_1) + W_1' \mathbf{x}_2 \\
& + \boldsymbol{\omega}_1 - \dot{\mathbf{y}}) - \varepsilon_1^T \tilde{W}_1 \phi(\mathbf{x}_1) \\
& + \varepsilon_1^T (-\lambda I_n \mathbf{e}_1 + \tilde{W}_1 \phi(\mathbf{x}_1)).
\end{aligned} \tag{4.8}$$

Now, define a virtual control $\mathbf{x}_2 = \alpha_1$ to force $\mathbf{z}_1 \to 0$, where

$$\alpha_1 = (W_1')^{-1}(\lambda \boldsymbol{\chi}_1 - W_1 \phi_1(\mathbf{x}_1) + \dot{y}_r - k_1 \mathbf{z}_1),$$

with $k_1 > 0$. Then, (4.8) becomes

$$\dot{V}(\mathbf{z}_1, \varepsilon_1, \tilde{W}_1) = -k_1 \|\mathbf{z}_1\|^2 - \lambda \|\varepsilon_1\|^2 + \mathbf{z}_1^T \boldsymbol{\omega}_1 + \mathbf{z}_1^T W_1' \mathbf{z}_2.$$

Step i: The error dynamics for $\mathbf{z}_j$ is derived from

$$\mathbf{z}_j = \mathbf{x}_j - \alpha_{j-1},$$

which represents the error between the actual control and the virtual control. Select the augmented Lyapunov function as

$$V(\mathbf{z}_1, \ldots, \mathbf{z}_j, \boldsymbol{\varepsilon}_1, \ldots, \boldsymbol{\varepsilon}_j, \tilde{W}_1, \ldots, \tilde{W}_j) =$$
$$V(\mathbf{z}_1, \ldots, \mathbf{z}_{j-1}, \boldsymbol{\varepsilon}_1, \ldots, \boldsymbol{\varepsilon}_{j-1}, \tilde{W}_1, \ldots, \tilde{W}_{j-1})$$
$$+ \frac{1}{2} \|\mathbf{z}_j\|^2 + \frac{1}{2} \|\boldsymbol{\varepsilon}_j\|^2 + \frac{1}{2\gamma} tr\left\{ \tilde{W}_j^T \tilde{W}_j \right\},$$

and its derivative is

$$\dot{V}(\mathbf{z}_1, \ldots, \mathbf{z}_j, \boldsymbol{\varepsilon}_1, \ldots, \boldsymbol{\varepsilon}_j, \tilde{W}_1, \ldots, \tilde{W}_j) =$$
$$\dot{V}(\mathbf{z}_1, \ldots, \mathbf{z}_{j-1}, \boldsymbol{\varepsilon}_1, \ldots, \boldsymbol{\varepsilon}_{j-1}, \tilde{W}_1, \ldots, \tilde{W}_{j-1})$$
$$+ \mathbf{z}_j^T \left(-\lambda \boldsymbol{\chi}_j + W_j \phi_j(\mathbf{x}_1, \ldots, \mathbf{x}_j) \right.$$
$$+ W_j' \mathbf{x}_{j+1} + \boldsymbol{\omega}_j - \dot{\alpha}_{j-1} \right) \tag{4.9}$$
$$+ \boldsymbol{\varepsilon}_j^T(-\lambda I_n \mathbf{e_j} + \tilde{W}_j \phi(\mathbf{x}_1, \ldots, \mathbf{x}_j))$$
$$- \boldsymbol{\varepsilon}_j^T \tilde{W}_j \phi(\mathbf{x}_1, \ldots, \mathbf{x}_j).$$

Analogously, the objective at the jth step is to select a virtual control $\mathbf{x}_{j+1} = \alpha_j$ to stabilize the error $\mathbf{z}_j = 0$, where

$$\alpha_j = (W_j')^{-1}(\lambda \boldsymbol{\chi}_j - W_j \phi_j(\mathbf{x}_1, \ldots, \mathbf{x}_j)$$
$$+ \dot{\alpha}_{j-1} - k_j \mathbf{z}_j), \tag{4.10}$$

with $k_j > 0$. Substituting (4.10) into (4.9) gives

$$\dot{V}(\mathbf{z}_1, \ldots, \mathbf{z}_j, \boldsymbol{\varepsilon}_1, \ldots, \boldsymbol{\varepsilon}_j, \tilde{W}_1, \ldots, \tilde{W}_j) =$$
$$- \sum_{k=1}^{j} (k_k \|\mathbf{z}_k\|^2 + \lambda \|\boldsymbol{\varepsilon}_k\|^2) + \sum_{k=1}^{j} \mathbf{z}_k^T \boldsymbol{\omega}_k + W_j' \mathbf{z}_j^T \mathbf{z}_{j+1}.$$

Step r: The sliding manifold is selected as $\mathbf{z}_r = x_r - \alpha_{r-1}$. The dynamics for system (4.5) in the **z**-variables are

$$\dot{z}_j = -W'_{j-1}z_{j-1} - k_j z_j + W'_j z_{j+1} + \omega_j,$$

$$\dot{z}_r = f_r(\mathbf{x}, \boldsymbol{\chi}, W) + W'_r u + \omega_r,$$

$$j = 1, \dots, r - 1.$$

Defining u as a discontinuous feedback controller of the form

$$u = (W'_r)^{-1} \left(-f_r(\mathbf{x}, \boldsymbol{\chi}, W)\right) - (W'_r)^{-1} k_r sign(\mathbf{z}_r)$$

$$= \bar{u}_{eq} - K_r\ sign\ (\mathbf{z}_r), \tag{4.11}$$

where $\bar{u}_{eq} = (W'_r)^{-1}\left(-f_r(\mathbf{x}, \boldsymbol{\chi}, W)\right)$ and $K_r = (W'_r)^{-1}k_r$. Define the Lyapunov function $V(\mathbf{z}_r, \boldsymbol{\varepsilon}_r, \tilde{W}_r) = \frac{1}{2}\mathbf{z}_r^T\mathbf{z}_r + \frac{1}{2}\boldsymbol{\varepsilon}_r^T\boldsymbol{\varepsilon}_r + \frac{1}{2\gamma}tr\left\{\tilde{W}_r^T\tilde{W}_r\right\}$ for the sliding variable, which has derivative

$$\dot{V}(\mathbf{z}_r, \boldsymbol{\varepsilon}_r, \tilde{W}_r) = \mathbf{z}_r^T\dot{\mathbf{z}}_r + \frac{1}{2}\boldsymbol{\varepsilon}_r^T\dot{\boldsymbol{\varepsilon}}_r + \frac{1}{2\gamma}tr\left\{\dot{\tilde{W}}_r^T\tilde{W}_r\right\}$$

$$= \mathbf{z}_r(-k_r\ sign\ (\mathbf{z}_r) + \omega_r) \tag{4.12}$$

$$\leq -k_r\|\mathbf{z}_r\|_1 - \lambda\|\boldsymbol{\varepsilon}_r\|_1 + \bar{\omega}_r\|\mathbf{z}_r\|_1,$$

where $k_r \geq u_0 + \bar{\omega}_r$ with $u_0 > 0$.

Using (4.12) and noting $\|\cdot\|_2 \leq \|\cdot\|_1$ [3], one obtains

$$\dot{V}(\mathbf{z}_r, \boldsymbol{\varepsilon}_r, \tilde{W}_r) \leq -u_0\|\mathbf{z}_r\|_2$$

$$\leq -u_0\sqrt{2}\left(V(\mathbf{z}_r, \boldsymbol{\varepsilon}_r, \tilde{W}_r)\right)^{1/2} < 0,$$

where $V(\mathbf{z}_r, \boldsymbol{\varepsilon}_r, \tilde{W}_r) > 0$ is positive definite. According to [1, Theorem 4.2] ($c = u_0\sqrt{2} > 0$, and $\alpha = \frac{1}{2}$), control law (4.11) guarantees the convergence of the pinned nodes to the manifold $\mathbf{z}_r = 0$ in a finite time defined by the settling time function as follows:

$$T(\mathbf{z}_r, \boldsymbol{\varepsilon}_r, \tilde{W}_r) \leq \frac{2}{u_0\sqrt{2}} V\left(\mathbf{z}_r, \boldsymbol{\varepsilon}_r, \tilde{W}_r\right)^{1/2}.$$

The sliding-mode equation for the pinned node is rewritten as

$$\dot{\mathbf{z}}_i = A_i \mathbf{z}_i + \boldsymbol{\omega}_i, \qquad i = 1, \ldots, l,$$

where $\mathbf{z}_{i_1} = [\mathbf{z}_{i_1}^T, \ldots, \mathbf{z}_{i_{r-1}}^T]^T$, $\boldsymbol{\omega}_i = [\boldsymbol{\omega}_{i_1}, \ldots, \boldsymbol{\omega}_{i_{r-1}}]^T$, and

$$A_i = \begin{bmatrix} -k_{i_1} & W'_{i_1} & 0 & \cdots & 0 \\ -W'_{i_1} & -k_{i_2} & W'_{i_2} & \cdots & 0 \\ \vdots & \ddots & \ddots & \ddots & \vdots \\ 0 & \cdots & 0 & W'_{i_{r-2}} & -k_{i_{r-1}} \end{bmatrix}.$$

Here, $\mathbf{z}_i$ can be considered as a linear stable variable, perturbed by the last terms $\boldsymbol{\omega}_i < \bar{\omega}_i$. Thus, the trajectories of $\mathbf{z}_i$ are ultimately bounded. The next condition is satisfied for the pinned nodes:

$$\frac{\partial V(\mathbf{z}_i)}{\partial \mathbf{z}_i} (A_i \mathbf{z}_i + \boldsymbol{\omega}_i) < (1 - \zeta)\sigma_{min}(A_i)\|\mathbf{z}_i\|^2,$$

$$\forall \|\mathbf{z}_i\| \geq \frac{\bar{\omega}_i}{\zeta\sigma_{min}(A_i)}.$$

For stability analysis of the whole network ($i = 1, 2, \ldots, N$), the following Lyapunov function is defined:

$$V_N(\mathbf{Z}) = \sum_{i=1}^{N} \frac{1}{2} \mathbf{z}_i^T \mathbf{P} \mathbf{z}_i, \tag{4.13}$$

$$\mathbf{Z} = [\mathbf{z}_1^T, \mathbf{z}_2^T, \ldots, \mathbf{z}_N^T]^T,$$

where $\mathbf{z}_i = \mathbf{x}_i - \boldsymbol{\alpha}_i$ and $\boldsymbol{\alpha}_i = [\mathbf{y}, \alpha_{i_1}, \ldots, \alpha_{i_{r-1}}]^T$. The time derivative of V_N

satisfies

$$
\begin{aligned}
\dot{V}_N(\mathbf{Z}) &= \sum_{i=1}^{N} \mathbf{z}_i^T \mathbf{P} \left(f_i(\mathbf{z}_i + \boldsymbol{\alpha}_i) - f_i(\boldsymbol{\alpha}_i) + \sum_{j=1}^{N} c_{ij} a_{ij} \mathbf{z}_j \right) \\
&< \sum_{i=1}^{N} \mathbf{z}_i^T \mathbf{P} \left(\sum_{j=1}^{N} c_{ij} a_{ij} \mathbf{z}_j - \psi_i \mathbf{z}_i \right) \\
&< \mathbf{Z}^T \left(-\boldsymbol{\psi} + \mathbf{G} \right) \otimes \mathbf{P} \mathbf{Z} < 0,
\end{aligned}
$$

$$
\forall \|\mathbf{z}_i\| \geq \frac{\bar{\omega}_i}{\zeta \sigma_{min}(A_i)},
$$

where $\psi_i = (1 - \zeta)\sigma_{min}(A_i)$ for the pinned nodes $(i = 1, 2, \ldots, l)$, ψ_i is given by Assumption 2 for the unpinned nodes $(i = l + 1, l + 2, \ldots, N)$ $\boldsymbol{\Psi} = diag(\psi_1, \psi_2, \ldots, \psi_N), \in \mathbb{R}^{N \times N}$, and $\mathbf{G} = (g_{ij}) = (c_{ij} a_{ij}) \in \mathbb{R}^{N \times N}$. The proof is thus completed. ☐

Remark 3. The performance of control law (4.6) can be improved by applying the boundary layer method [10], which removes chattering in the SMC. The boundary layer method replaces the signum function $sign(\cdot)$ by a saturation function $sat(\cdot)$.

4.4 Simulation Results

To illustrate the applicability of the control scheme, a directed complex network (4.3), consisting of 10 nodes with 3 different self-dynamics, is simulated, with a topological structure given by

$$A = \begin{bmatrix} -7 & 1 & 1 & 1 & 0 & 1 & 1 & 0 & 1 & 1 \\ 1 & -5 & 1 & 1 & 0 & 0 & 0 & 1 & 1 & 0 \\ 1 & 1 & -4 & 0 & 1 & 0 & 1 & 0 & 0 & 0 \\ 1 & 1 & 0 & -3 & 0 & 0 & 1 & 0 & 0 & 0 \\ 0 & 0 & 1 & 0 & -1 & 0 & 0 & 0 & 0 & 0 \\ 1 & 0 & 0 & 0 & 0 & -1 & 0 & 0 & 0 & 0 \\ 0 & 0 & 0 & 1 & 0 & 0 & -1 & 0 & 0 & 0 \\ 0 & 1 & 0 & 0 & 0 & 0 & 0 & -1 & 0 & 0 \\ 0 & 1 & 0 & 0 & 0 & 0 & 0 & 0 & -1 & 0 \\ 1 & 0 & 0 & 0 & 0 & 0 & 0 & 0 & 0 & -1 \end{bmatrix}.$$

Nodes 1–4 are described by Chua's circuit [7] . Nodes 5–7 are selected as a nonlinear system described by

$$\dot{x}_1 = -x_1$$
$$\dot{x}_2 = -x_1 - x_2 - x_3 - x_1 x_3 \qquad (4.14)$$
$$\dot{x}_3 = x_2(x_1 + 1).$$

Nodes 8–10 are described by an unstable linear system,

$$\dot{x}_1 = -x_1 + x_2$$
$$\dot{x}_2 = x_2 + x_3 \qquad (4.15)$$
$$\dot{x}_3 = -x_2 - 1.5x_3,$$

where c_{ij} are selected as constants equal to 50. The output signal $y(t)$ is selected as

$$y(t) = 0.5 \sin\left(\frac{\pi}{4}t\right).$$

The passivity degrees defined in Assumption 2 for the three types of nodes are $\psi_{Chua} = -2$ (Chua's circuit (2.11)), $\psi_{NS} = 0.2$ (nonlinear system (4.14)), $\psi_{LS} = -1$ (unstable linear system (4.15)), which are obtained via simulations.

The matrix M in condition (4.7) without controller has one non-negative eigenvalue. Furthermore, based on [11, Proposition 7], one controller is required to achieve output synchronization. For this case, Node 1 is selected as the pinned one. The neural identifier (4.4) is given by

$$\dot{\chi}_{1_1} = -16\chi_{1_1} + W_1\phi_{1_1}(x_{1_1}) + x_{1_2}$$
$$\dot{\chi}_{1_2} = -16\chi_{1_2} + W_2\phi_{1_2}(x_{1_1}, x_{1_2}) + x_{1_3}$$
$$\dot{\chi}_{1_3} = -16\chi_{1_3} + W_3\phi_{1_3}(\mathbf{x}_1) + u,$$

with

$$\phi_{1_1}(x_{1_1}) = [\mathcal{S}(x_{1_1}); \mathcal{S}^2(x_{1_2}); \mathcal{S}^4(x_{1_1})],$$
$$\phi_{1_2}(x_{1_1}, x_{1_2}) = [\mathcal{S}(x_{1_1}); \mathcal{S}(x_{1_2});$$
$$\mathcal{S}(x_{1_1})\mathcal{S}(x_{1_2}); \mathcal{S}^2(x_{1_1}); \mathcal{S}^2(x_{1_2})],$$
$$\phi_{1_3}(\mathbf{x}_1) = [\mathcal{S}(x_{1_1})^2; \mathcal{S}(x_{1_2})^2;$$
$$\mathcal{S}(x_{1_1})\mathcal{S}(x_{1_2}); \mathcal{S}^4(x_{1_1}); \mathcal{S}^4(x_{1_2})].$$

The controller is defined as (4.6), which fulfills condition (4.7).

The equation of the pinned node $\mathbf{X}_1$ is given by

$$\dot{x}_{1_1} = P_C * (G_C(x_{1_2} - x_{1_1}) - \phi(x_{1_1})) + 50\sum_{j=1}^{10} a_{1j}x_{j_1}$$

$$\dot{x}_{1_2} = G_C(x_{1_1} - x_{1_2}) + x_{1_3} + 50\sum_{j=1}^{10} a_{1j}x_{j_2} \qquad (4.16)$$

$$\dot{x}_{1_3} = -Q_C x_{1_2} + 50\sum_{j=1}^{10} a_{1j}x_{j_3} + u.$$

Initially, $c_{ij} = 0$ and $u = 0$ are selected. At $t = 2s$, the network is connected with the designed coupling strengths. Then, at $t = 2.2s$, the controller is switched on, and output synchronization is achieved. Fig. 4.2 shows the state evolution of the whole network. Fig. 4.3 presents the control input signal $u_i(t)$ applied to the pinned node. It is easy to see that no chattering occurs in the control signal, thanks to the boundary layer method mentioned in Remark 3. Fig. 4.4 displays the synchronization error $\bar{z}_1$. Finally, Figs. 4.5 and 4.6 portray the evolutions of identification errors ε_{i_i} and weights for Node 1, respectively, which are bounded, illustrating good neural identifier performance.

Table 4.1 presents statistical measures (Mean Square Error and Standard Deviation) of the neural sliding-mode pinning controller (the identification error ε_{i_i}, the tracking error for the pinned node (Node 1) z_1, and the average tracking error $\bar{z}_1$).

TABLE 4.1

Statistical measures of the neural sliding-mode pinning controller.

Measure	ε_{i_i}	z_1	$\bar{z}_1$
Mean Square Error	0.6545	1.0161×10^{-4}	1.1053×10^{-4}
Standard Deviation	0.8084	0.0101	0.0105

By these simulation results, it is verified that the neural sliding-mode pinning control scheme achieves output synchronization successfully for the whole complex network.

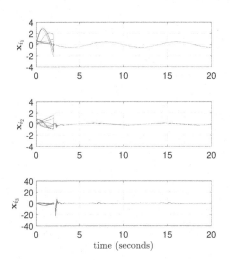

FIGURE 4.2
Evolution of network states.

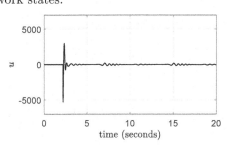

FIGURE 4.3
Control input u for the pinned node.

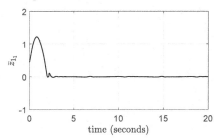

FIGURE 4.4
Average output synchronization error $\bar{z}_1$.

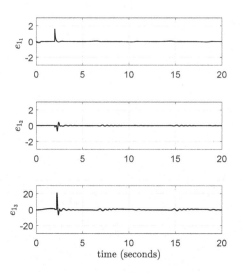

FIGURE 4.5
Identification of error ε_{i_i} for the pinned node.

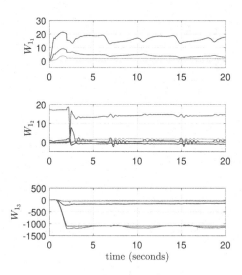

FIGURE 4.6
Weights evolution at Node 1 with identification.

4.5 Conclusions

This chapter has established a neural sliding-mode pinning control strategy to achieve output synchronization for uncertain general complex networks. This controller is composed by two components: an RHONN identifier and a sliding-mode controller for output synchronization. For determining the sliding manifold, the backstepping technique is applied. Simulation results are presented to illustrate the effectiveness of the proposed control scheme in a directed complex network with non-identical nodes and unknown dynamics.

Bibliography

[1] S. Bhat and D. Bernstein. Finite-time stability of continuous autonomous systems. *SIAM Journal on Control and Optimization*, 38(3):751–766, 2000.

[2] G. Chen. Pinning control and controllability of complex dynamical networks. *International Journal of Automation and Computing*, 1–9, 2017.

[3] H. Khalil. *Nonlinear Systems.* Upper Saddle River, NJ, USA: Prentice-Hall, 2002.

[4] E. Kosmatopoulos, M. Christodoulou, and P. Ioannou. Dynamical neural networks that ensure exponential identification error convergence. *Neural Networks*, 10(2):299–314, 1997.

[5] M. Krstic, I. Kanellakopoulos, and P. Kokotovic. *Nonlinear and Adaptive Control Design.* Wiley, 1995.

[6] X. Li, X. Wang, and G. Chen. Pinning a complex dynamical network to its equilibrium. *IEEE Transactions on Circuits and Systems I: Regular Papers*, 51(10):2074–2087, 2004.

[7] T. Matsumoto, L. Chua, and M. Komuro. The double scroll. *IEEE Transactions on Circuits and Systems*, 32(8):797–818, 1985.

[8] G. Rovithakis and M. Christodoulou. *Adaptive Control with Recurrent High-order Neural Networks: Theory and Industrial Applications.* London, UK: Springer-Verlag, 2000.

[9] L. Ricalde and E. Sanchez. Output tracking with constrained inputs via inverse optimal adaptive recurrent neural control. *Engineering Applications of Artificial Intelligence*, 21(4):591–603, 2008.

[10] J. Slotine and W. Li. *Applied Nonlinear Control*. Englewood Cliffs, NJ, USA: Prentice Hall, 1991.

[11] J. Xiang and G. Chen. On the V-stability of complex dynamical networks. *Automatica*, 43(6):1049–1057, 2007.

Part III

Optimal Control

5

Model-Based Optimal Control

In this chapter, a novel approach to trajectory tracking control of a complex network with identical and non-identical nodes is presented, based on the idea and technique of inverse optimal pinning control. A control law, which stabilizes the tracking error dynamics is synthesized. This controller manages different coupling strengths, and simultaneously minimizes an associated cost functional. Chaotic systems are used to illustrate the applicability of the methodology and the controller, as well as their effectiveness, via simulations.

5.1 Trajectory Tracking of Complex Networks

This section develops an optimal pinning control scheme for solving trajectory tracking of complex network, as illustrated by Fig. 5.1. The control objective is to synthesize a controller, which can force (2.14) to track the state of a reference system given by

$$\dot{\mathbf{x}}_r = f_r(\mathbf{x}_r), \quad \mathbf{x}_r \in \mathbb{R}^n, \tag{5.1}$$

i.e., the tracking error $\mathbf{e}_i = \mathbf{x}_i - \mathbf{x}_r \in \mathbb{R}^n$ satisfies

$$\lim_{t \to \infty} \|\mathbf{e}_i(t)\| = 0, \quad \forall\, i = 1, 2, \ldots, N, \tag{5.2}$$

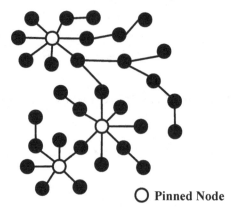

(a) Pinned Controlled Complex Network Scheme

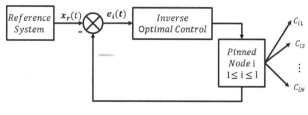

(b) Pinned Node

FIGURE 5.1
The proposed control scheme.

and simultaneously minimizes an associated cost functional defined by

$$J_i = \lim_{t \to \infty} \left[\mathfrak{J}(\mathbf{e}_i(t)) + \int_0^t \left(\mathfrak{l}(\mathbf{e}_i) + \mathbf{u}_i^T R(\mathbf{e}_i) \mathbf{u}_i \right) d\tau \right], \tag{5.3}$$

where $R = R^T > 0$ for all $\mathbf{e}_i = \mathbf{x}_i - \mathbf{x}_r \in \mathbb{R}^n$, and $\mathfrak{l}(\mathbf{e}_i)$ and $\mathfrak{J}(\mathbf{e}_i)$ are positive definite and radially unbounded functions [2]. The control strategy is applied only to a fraction of nodes (pinned ones). Suppose that the node dynamics are known. The main result is the following theorem.

Theorem 5.1. *Consider the complex network with pinning control (2.14), in which $\Gamma = \mathbf{I}_n$, where a reference system is defined by (5.1), with all $f(\mathbf{x}_i)$, $f_r(\mathbf{x}_r)$ being chaotic systems. If c_{min}, the minimal coupling strength of the* *entire network, satisfies*

$$c_{min} = \beta > \frac{h_{max}}{\sigma_{min}(-A + diag(\hat{d}_{k_1}, \hat{d}_{k_2}, ..., \hat{d}_{k_l}, 0, ..., 0))}, \quad (5.4)$$

where h_{max} is the maximum positive Lyapunov exponent of the error systems $(f(\mathbf{x}_i) - f_r(\mathbf{x}_r))$, for all $i = 1, 2, ..., N$, and

$$\hat{d}_{k_i} = \begin{cases} \frac{\varpi + \sqrt{\varpi^2 + \left(\mathbf{e}_i^T \mathbf{e}_i\right)^2}}{\mathbf{e}_i^T \mathbf{e}_i} & \|\mathbf{e}_i\| \neq 0 \\ 0 & \|\mathbf{e}_i\| = 0, \end{cases} \quad (5.5)$$

with

$$\varpi = \mathbf{e}_i^T \left(f(\mathbf{x}_i) - f_r(\mathbf{x}_r)\right) + k_b \|\mathbf{e}_i\|^2,$$

then the optimal pinning control law

$$\mathbf{u}_i = \alpha_i^*(\mathbf{e}_i), \quad i = 1, 2, ..., l, \quad (5.6)$$

with

$$\alpha_i^*(\mathbf{e}_i) = -\beta R_2^{-1}(L_{g_2}V)^T = -c_{min}\hat{d}_{k_i}\mathbf{e}_i, \quad (5.7)$$

where $k_b > 2$ is a design parameter, guarantees that the whole complex network tracks the desired reference. Furthermore, $\mathbf{u}_i$ minimizes the cost functional

$$J(u) = \sup_{d \in D} \left\{ \lim_{t \to \infty} \left[\|\mathbf{e}_i\|^2 + \int_0^t \left(-\beta\lambda \|\mathbf{e}_i\|^2 + 2\alpha_i^T R_2(\mathbf{e}_i)\alpha_i - \frac{\beta|\omega|}{\lambda^2} \right) d\tau \right] \right\}, \quad (5.8)$$

for each pinned node, where

$$\omega = \sum_{j=1}^N c_{ij} a_{ij} \Gamma \mathbf{x}_j, \quad \beta \geq 2, \quad \lambda \in (0, 2], \quad R_2(\mathbf{e}_i) = \frac{1}{\hat{d}_{k_i}}.$$

Proof. The tracking error $\mathbf{e}_i$ is given by

$$\dot{\mathbf{e}}_i = f(\mathbf{x}_i) - f_r(\mathbf{x}_r) + \sum_{j=1}^N c_{ij} a_{ij} \Gamma \mathbf{x}_j + \mathbf{u}_i, \quad i = 1, 2, ..., l. \quad (5.9)$$

Then, a control Lyapunov function is chosen as

$$V = \frac{1}{2}\mathbf{e}_i{}^T\mathbf{e}_i. \tag{5.10}$$

By taking its time derivative, one obtains

$$
\begin{aligned}
\dot{V} &= L_fV + L_{g_1}V\boldsymbol{\omega} + L_{g_2}V\mathbf{u}_i = \mathbf{e}_i^T\dot{\mathbf{e}}_i \\
&= \mathbf{e}_i^T\left(f(\mathbf{x}_i) - f_r(\mathbf{x}_r) + \sum_{j=1}^{N}c_{ij}a_{ij}\boldsymbol{\Gamma}\mathbf{x}_j + \mathbf{u}_i\right),
\end{aligned}
$$

where

$$
\begin{aligned}
L_fV &= \mathbf{e}_i{}^T\left(f(\mathbf{x}_i) - f_r(\mathbf{x}_r)\right), \\
L_{g_1}V &= \mathbf{e}_i{}^T, \\
L_{g_2}V &= \mathbf{e}_i{}^T.
\end{aligned}
$$

From Theorem 2.9, auxiliary system (2.31) for (5.9) is given by

$$\dot{\mathbf{e}}_i = f(\mathbf{x}_i) - f_r(\mathbf{x}_r) + \frac{\ell\gamma\left(2\left\|\mathbf{e}_i\right\|\right)}{\left\|\mathbf{e}_i\right\|^2}\mathbf{e}_i + \mathbf{u}_i, \tag{5.11}$$

where $f(x) = f(\mathbf{e}_i) - f_r(\mathbf{x}_r)$, $g_1(x) = I_n$, $g_2(x) = I_n$. Based on (5.11), the control input α_i is chosen as (2.35), (2.36), with

$$
\begin{aligned}
\boldsymbol{\varpi} &= L_fV + |L_{g_1}V|\rho^{-1}\left(\|\mathbf{e}_i\|\right) \\
&= \mathbf{e}_i{}^T\left(f(\mathbf{e}_i) - f_r(\mathbf{x}_r)\right) + k_b\|\mathbf{e}_i\|^2, \tag{5.12}
\end{aligned}
$$

and

$$\rho(r) = \frac{1}{k_b}r, \quad k_b > 2. \tag{5.13}$$

The optimal control law $\mathbf{u}_i = \alpha_i{}^*(\mathbf{e}_i)$ is defined as

$$
\begin{aligned}
\alpha_i{}^*(\mathbf{e}_i) &= -\beta R_2{}^{-1}(L_{g_2}V)^T \\
&= -\beta\left(\frac{\varpi + \sqrt{\varpi^2 + (\mathbf{e}_i{}^T\mathbf{e}_i)^2}}{\mathbf{e}_i{}^T\mathbf{e}_i}\right)\mathbf{e}_i, \tag{5.14}
\end{aligned}
$$

where

$$R_2(\mathbf{e}_i) = \begin{cases} \dfrac{\mathbf{e}_i{}^T\mathbf{e}_i}{\varpi + \sqrt{\varpi^2 + (\mathbf{e}_i{}^T\mathbf{e}_i)^2}} > 0, & \|\mathbf{e}_i\| \neq 0 \\[2ex] k_b, & \|\mathbf{e}_i\| = 0. \end{cases} \tag{5.15}$$

To calculate $\ell_\gamma\left(2\left\|\mathbf{e}_i\right\|\right)$, the Legendre-Fenchel transform of $\gamma\left(\left\|\mathbf{e}_i\right\|\right) = \left\|\mathbf{e}_i\right\|^2 \in \mathcal{K}_\infty$ is used [2], where $\left(\gamma'\left(\left\|\mathbf{e}_i\right\|\right)\right)^{-1} = \frac{1}{2}\left\|\mathbf{e}_i\right\| \in \mathcal{K}_\infty$, $\gamma'\left(\left\|\mathbf{e}_i\right\|\right) = 2\left\|\mathbf{e}_i\right\| \in \mathcal{K}_\infty$, so that $\ell_\gamma\left(2\left\|\mathbf{e}_i\right\|\right) = \left\|\mathbf{e}_i\right\|^2$. As a result, system (5.11) can be rewritten as

$$\dot{\mathbf{e}}_i = f(\mathbf{e}_i) - f_r(\mathbf{x}_r) + \mathbf{e}_i - \left(\frac{\varpi + \sqrt{\varpi^2 + (\mathbf{e}_i{}^T\mathbf{e}_i)^2}}{\mathbf{e}_i{}^T\mathbf{e}_i}\right)\mathbf{e}_i. \tag{5.16}$$

Consequently, the derivate of the Lyapunov function (5.10), along the solution of (5.16), is obtained as

$$\begin{aligned} \dot{V}\Big|_{u=\frac{\alpha_i}{2}} &= \mathbf{e}_i{}^T\left[f(\mathbf{x}_i) - f_r(\mathbf{x}_r) + \mathbf{e}_i - \frac{1}{2}\left(\frac{\overbrace{\varpi + \sqrt{\varpi^2 + (\mathbf{e}_i{}^T\mathbf{e}_i)^2}}^{\varpi}}{\mathbf{e}_i{}^T\mathbf{e}_i}\right)\mathbf{e}_i \right] \\ &= \mathbf{e}_i{}^T\left(f(\mathbf{x}_i) - f_r(\mathbf{x}_r)\right) + k_b\|\mathbf{e}_i\|^2 + \|\mathbf{e}_i\|^2 \\ &\quad -\frac{1}{2}\left(\varpi + \sqrt{\varpi^2 + (\mathbf{e}_i{}^T\mathbf{e}_i)^2}\right) - k_b\|\mathbf{e}_i\|^2 \\ &= \varpi + \|\mathbf{e}_i\|^2 - \frac{1}{2}\left(\varpi + \sqrt{\varpi^2 + (\mathbf{e}_i{}^T\mathbf{e}_i)^2}\right) - k_b\|\mathbf{e}_i\|^2 \\ &= \frac{1}{2}\left(\varpi - \sqrt{\varpi^2 + (\mathbf{e}_i{}^T\mathbf{e}_i)^2}\right) - (k_b - 1)\|\mathbf{e}_i\|^2 \\ &= -W(\mathbf{e}_i) - (k_b - 1)\|\mathbf{e}_i\|^2 < 0, \end{aligned}$$

where

$$W(\mathbf{e}_i) = -\frac{1}{2}\left(\varpi - \sqrt{\varpi^2 + (\mathbf{e}_i{}^T\mathbf{e}_i)^2}\right) > 0. \tag{5.17}$$

As a result, (5.11) is asymptotically stabilized.

Once stability is achieved, using control law (5.14), it is to show that the cost functional (2.33) is minimized. To do so, $l(\mathbf{e}_i)$ must be positive definite and radially unbounded. From (2.34), it follows that

$$
\begin{aligned}
l(\mathbf{e}_i) &= -2\beta L_f V - \lambda\beta\|\mathbf{e}_i\|^2 + \beta\left(\varpi + \sqrt{\varpi^2 + (\mathbf{e}_i^T\mathbf{e}_i)^2}\right) \\
&= -2\beta\mathbf{e}_i^T\left(f(\mathbf{x}_i) - f_r(\mathbf{x}_r)\right) - \beta k_b\|\mathbf{e}_i\|^2 + \beta k_b\|\mathbf{e}_i\|^2 \\
&\quad + \beta\left(\varpi + \sqrt{\varpi^2 + (\mathbf{e}_i^T\mathbf{e}_i)^2}\right) - \lambda\beta\|\mathbf{e}_i\|^2 \\
&= \beta\left(-\varpi + \sqrt{\varpi^2 + (\mathbf{e}_i^T\mathbf{e}_i)^2}\right) + \beta\left(k_b - \lambda\right)\|\mathbf{e}_i\|^2. \qquad (5.18)
\end{aligned}
$$

Placing (5.17) in (5.18), one obtains

$$
l(\mathbf{e}_i) = 2\beta W(\mathbf{e}_i) + \beta\left(k_b - \lambda\right)\|\mathbf{e}_i\|^2 > 0, \quad \forall\lambda \in (0,2].
$$

It is easy to see that $l(\mathbf{e}_i) \to \infty$ as $\|\mathbf{e}_i\| \to \infty$; hence, $l(\mathbf{e}_i)$ is positive definite and radially unbounded.

Then, from Theorem 2.9, the control law (5.14) minimizes the cost functional

$$
J(u) = \sup_{d\in D}\left\{\lim_{t\to\infty}\left[\|\mathbf{e}_i\|^2 + \int_0^t\left(-\beta\lambda\|\mathbf{e}_i\|^2 + 2\alpha_i^T R_2(\mathbf{e}_i)\alpha_i - \tfrac{\beta|\omega|}{\lambda^2}\right)d\tau\right]\right\}
$$

with $\beta \geq 2$, and the value function of (5.3) is $J^*(\mathbf{e}_i) = \beta\|\mathbf{e}_i\|^2$, which is unbounded.

Now, the stability of tracking error for the whole network is analyzed, following the same procedure as in [4]. The controlled network (2.14) is linearized at $\mathbf{e}_i = 0$ component-wise, with $\mathbf{u}_i$ as given in (5.7), so that

$$
\dot{\eta} = \eta\left[D\mathbf{f}(\mathbf{x}_i, \mathbf{x}_r)\right] - \hat{B}\eta,
$$

where $D\mathbf{f}(\mathbf{x}_i, \mathbf{x}_r) \in \mathbb{R}^{n\times n}$ is the Jacobian of

$$
\mathbf{f}(\mathbf{x}_i, \mathbf{x}_r) = f(\mathbf{x}_i) - f_r(\mathbf{x}_r), \quad i = 1, 2, \ldots, N, \qquad (5.19)
$$

with $\eta = (\eta_1, \eta_2, \ldots, \eta_N)^T \in \mathbb{R}^{Nn}$, $\eta_i(t) = x_i(t) - x_r(t)$, $i = 1, 2, \ldots, N$, and $\hat{B} = G + diag(c_{\min}\hat{d}_{k_1}, \ldots, c_{\min}\hat{d}_{k_l}, 0, \ldots, 0)$, in which $\hat{d}_{k_i} = R_2^{-1}(\mathbf{e_i})$ as in (5.15). Note that $\hat{B}$ is symmetric and positive definite; hence, all of its eigenvalues are positive.

According to [8], one has

$$
\begin{aligned}
0 \ <\ & \sigma_{\min}\left(c_{\min}\left[-A + diag(\hat{d}_{k_1}, \ldots, \hat{d}_{k_l}, 0, \ldots, 0)\right]\right) \\
\leq\ & \sigma_{\min}\left(G + diag(c_{\min}\hat{d}_{k_1}, \ldots, c_{\min}\hat{d}_{k_l}, 0, \ldots, 0)\right).
\end{aligned}
$$

This inequality is obtained based on the fact that $c_{ij} \geq c_{min} > 0$ for all c_{ij} in G. Then, the coupling strengths can be different for different nodes.

Furthermore, the Transversal Lyapunov Exponents (TLEs) denoted by $\mu_k(\sigma_i)$, for each eigenvalue σ_i, $i = 1, 2, \ldots, N$, are given by $\mu_k(\sigma_i) = h_k - c\sigma_i$, $k = 1, 2, \ldots n$ [3], which determine the stability of the controlled states. Hence, local stability of the controlled network (2.14) requires negative TLEs. In order to guarantee that TLEs are negative, the following condition must be satisfied:

$$
\mu_{\max}(\sigma_{\min}) = h_{\max} - c_{\min}\sigma_{min}(-A + diag(\hat{d}_{k_1}, \hat{d}_{k_2}, \ldots, \hat{d}_{k_l}, 0, \ldots, 0)) < 0,
$$

which is equivalent to condition (5.4). Then, with the control law (5.6), trajectory tracking is achieved for the whole complex network. $\square$

Simulations are performed using Matlab/Simulink[1]. To illustrate the performance of the proposed scheme, a scale-free network composing of 50 nodes is simulated, where only one node is pinned. Here, the node with the highest degree is pinned, with each network node selected as the chaotic Chen's system [1]. Suppose that $\Theta = diag(1, 1, 1)$. Define the state variables

---

[1]MATLAB & Simulink are registered trademarks of MathWorks Inc., Natick, Massachusetts, U.S.A.

as: $x_1 = x$, $x_2 = y$, $x_3 = z$. The equation of the pinned node $\mathbf{x}_1$ is

$$\dot{x}_{1_1} = a_C(x_{1_2} - x_{1_1}) + \sum_{j=1}^{50} c_{1j}a_{1j}x_{j_1} + u_{1_1}$$

$$\dot{x}_{1_2} = (c_C - a_C)x_{1_1} - x_{1_1}x_{1_3} + c_C x_{1_2} + \sum_{j=1}^{50} c_{1j}a_{1j}x_{j_2} + u_{1_2} \quad (5.20)$$

$$\dot{x}_{1_3} = x_{1_1}x_{1_2} - b_C x_{1_3} + \sum_{j=1}^{50} c_{1j}a_{1j}x_{j_3} + u_{1_3}.$$

Simulations are performed as follows: From $t = 0s$ to $t = 3.8s$, the network runs without any connection, i.e., $(c = 0)$. At $t = 3.8s$, the coupling strengths are set constant, randomly chosen with $c_{ij} \geq 200$ for 80% of nodes of the entire network, and the values of coupling strengths c_{ij} are time-varying for the other nodes. Then, at $t = 4s$, the control law is applied, consequently, the network is stabilized at a constant reference, which is selected as the unstable equilibrium point, $x_r = [7.9373, 7.9373, 21]^T$. Afterwards, at $t = 8s$, a reference signal selected as the chaotic Rössler system [7] is incepted to generate the desired trajectory.

Fig. 5.2(a) presents the results of state-time evolution and Fig. 5.2(b) displays the respective phase portrait for the pinning node (solid line) and reference (dash-dot line). As can be seen, tracking is achieved for the pinning node. Fig. 5.3(a) presents the results for the control input signal $u_i(t)$ applied to the pinned node. Furthermore, Fig. 5.3(b) shows the tracking error e_{t_i} for the pinning node. It is easy to see that the control scheme drives the whole network to track the desired values, which confirm its effectiveness. Fig. 5.3(c) presents the values of coupling strengths c_{ij}. Finally, Fig. 5.3(d) shows the average trajectory of the whole network. From these results, it is verified that the optimal pinning control scheme achieves trajectory tracking for the whole complex network.

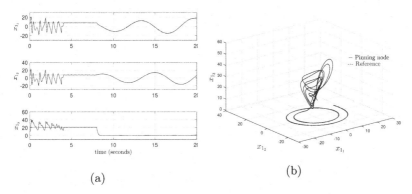

FIGURE 5.2
Pinning node $x_{1_i}(t)$, $i = 1, 2, 3$: (a) Time evolution, (b) Phase portrait.

5.2 Non-Identical Nodes

This section presents an inverse optimal pinning control scheme for solving trajectory tracking of complex network based on V-stability theory. The main result is established in the following lemma.

Lemma 5.1. *Consider the complex network (2.21), with reference system defined by (5.1). Suppose that condition*

$$\mathbf{M} = (-\mathbf{\Theta} + \mathbf{G} - \mathbf{\Psi}) \otimes \mathbf{P}\mathbf{\Gamma} < 0 \qquad (5.21)$$

is satisfied, where $\mathbf{\Theta} = diag(\theta_1, \theta_2, \ldots, \theta_N) \in \mathbb{R}^{N \times N}$, $\mathbf{G} = (g_{ij}) = (c_{ij} a_{ij}) \in \mathbb{R}^{N \times N}$, $\mathbf{\Psi} \in \mathbb{R}^{N \times N}$ is a diagonal matrix with the first l elements $\psi_i \neq 0$, $i = 1, 2, \ldots, l$, and all the other $(N-l)$ elements are equal to zero, $\mathbf{P} = \mathbf{P}^T > 0$, and the control law

$$\mathbf{u}_i = -\frac{1}{2}R^{-1}(\mathbf{e}_i)L_g V^T = \alpha_i^*(\mathbf{e}_i), \qquad (5.22)$$

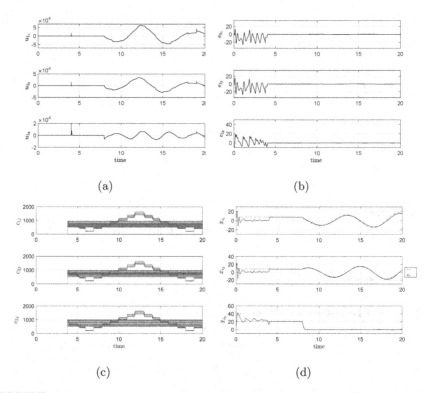

FIGURE 5.3
Results for a scale-free network composing of 50 nodes, where each network node selected as the chaotic Chen's system: (a) Optimal pinning control law $u_{1_i}(t)$, $i = 1, 2, 3$; (b) Tracking error $\mathbf{e}_{1_i}$, $i = 1, 2, 3$; (c) Coupling strengths c_{ij}, $i = 1, 2, 3$, $j = 1, 2, \ldots, 50$; (d) Average trajectory of the whole network.

where

$$\alpha_i^* = \begin{cases} -\left[\dfrac{\varpi + \sqrt{\varpi^2 + \left(L_g V (L_g V)^T\right)^2}}{L_g V (L_g V)^T}\right] L_g V^T, & \|L_g V\| \neq 0, \\[4mm] 0, & \|L_g V\| = 0, \end{cases}$$

in which

$$\varpi = L_f V + |L_h V| \, \rho^{-1}\left(\|\mathbf{e}_i\|\right), \tag{5.23}$$

where $\rho \in \mathcal{K}_\infty$ for all $\mathbf{e}_i \neq 0$, and fulfills $\|\mathbf{e}_i\| \geq \rho(\|\boldsymbol{\omega}\|)$ with

$$\|\boldsymbol{\omega}\| = \left\| \sum_{j=1}^{N} c_{ij} a_{ij} \boldsymbol{\Gamma} \mathbf{x}_j \right\|.$$

Then, control law (5.22) guarantees that the whole network tracks the desired reference, simultaneously minimizing (5.3).

Proof. Tracking error stability for the whole network is analyzed. Consider the following Lyapunov function for the whole controlled network (2.21):

$$V_N(\mathbf{E}) = \sum_{i=1}^{N} \frac{1}{2} \mathbf{e}_i^T \mathbf{P} \mathbf{e}_i, \tag{5.24}$$

$$\mathbf{E} = [\mathbf{e}_1^T, \mathbf{e}_2^T, \ldots, \mathbf{e}_N^T]^T.$$

By taking its time derivative, one obtains

$$\dot{V}_N(\mathbf{E}) = \sum_{i=1}^{N} \mathbf{e}_i^T \mathbf{P} \left(f_i(\mathbf{x}_i) - f_r(\mathbf{x}_r) + \sum_{j=1}^{N} c_{ij} a_{ij} \boldsymbol{\Gamma} \mathbf{e}_j + \alpha_i(\mathbf{e}_i) \right)$$

$$< \sum_{i=1}^{N} \mathbf{e}_i^T \mathbf{P} \left(\sum_{j=1}^{N} c_{ij} a_{ij} \boldsymbol{\Gamma} \mathbf{e}_j - (\theta_i + \psi_i) \boldsymbol{\Gamma} \mathbf{e}_i \right)$$

$$< \mathbf{E}^T (-\boldsymbol{\Theta} + \mathbf{G} - \boldsymbol{\Psi}) \otimes \mathbf{P} \boldsymbol{\Gamma} \mathbf{E} < 0, \tag{5.25}$$

where $\boldsymbol{\Theta} = diag(\theta_1, \theta_2, \ldots, \theta_N) \in \mathbb{R}^{N \times N}$, $\mathbf{G} = (g_{ij}) = (c_{ij} a_{ij}) \in \mathbb{R}^{N \times N}$, $\boldsymbol{\Psi} \in \mathbb{R}^{N \times N}$ is a diagonal matrix with the first l elements ψ_i, $i = 1, 2, \ldots, l$, while other $(N - l)$ elements are all zero. Inequality (5.25) is satisfied. Then, following the same procedure as in Theorem 5.4, it can be proved that (5.22) is an inverse optimal control law that minimizes (5.3) for the pinned nodes. $\square$

Lemma 5.1 is valid for complex networks with non-identical nodes. Additionally, Equation (5.21) facilitates the analysis of different characteristics associated with a complex network, such as stability, robustness, and selection of the pinned nodes number, by studying one simple matrix that characterizes

the network topology. This is a great advantage of the V-stability approach
[9]. Here, it is assumed that all the nodes have the same dimension, and their
dynamics are known.

To verify the applicability of this controller, consider a heterogeneous
network composing of 9 nodes with three types of chaotic nodes, described by
Chen's system (Nodes 1, 2, and 3) [1], Lorenz's system (Nodes 4, 5, and 6)
[5], and Lü's system (Nodes 7, 8, and 9) [6], which are three-dimensional
autonomous chaotic systems with different complex dynamical behaviors.
The coupling strengths are constants $c_{ij} = c = 100$, and $\mathbf{\Gamma} = diag(1,1,1)$.
Fig. 5.4(a) illustrates this network. The control goal is to track a reference
signal generated by the Rössler system (2.12). To verify that condition
(5.21) is fulfilled, consider the Lyapunov function $V_N(\mathbf{E}) = \sum_{i=1}^{N} \frac{1}{2}\mathbf{e}_i^T P\mathbf{e}_i$,
with $\mathbf{P} = diag(1,1,1)$. By simulations, it is found that there exist passivity
degrees given by (2.22) for the Chen system, Lorenz system, and Lü system,
respectively, when the reference trajectory is from the Rössler system, with
$\theta_{CR} = -80$, $\theta_{LR} = -60$, and $\theta_{LuR} = -70$, respectively. The matrix $\mathbf{A}$ for the
network is

$$
A = \begin{bmatrix}
-7 & 1 & 1 & 1 & 1 & 1 & 1 & 1 & 0 \\
1 & -4 & 1 & 1 & 0 & 0 & 1 & 0 & 0 \\
1 & 1 & -6 & 0 & 1 & 1 & 0 & 1 & 1 \\
1 & 1 & 0 & -2 & 0 & 0 & 0 & 0 & 0 \\
1 & 0 & 1 & 0 & -2 & 0 & 0 & 0 & 0 \\
1 & 0 & 1 & 0 & 0 & -2 & 0 & 0 & 0 \\
1 & 1 & 0 & 0 & 0 & 0 & -3 & 0 & 1 \\
1 & 0 & 1 & 0 & 0 & 0 & 0 & -2 & 0 \\
0 & 0 & 1 & 0 & 0 & 0 & 1 & 0 & -2
\end{bmatrix}.
$$

For $\mathbf{\Psi} = 0_N$, the open-loop system matrix $\mathbf{M}$ in condition (5.21) has one
non-negative eigenvalue. By [9, Proposition 7], one controller is sufficient to
ensure trajectory tracking. The node with the highest degree is selected as

pinned node [4]. Then, the controller is defined as (5.22), i.e.,

$$\mathbf{u}_1 = -\left(\frac{\varpi + \sqrt{\varpi^2 + (\mathbf{e}_1{}^T\mathbf{e}_1)^2}}{\mathbf{e}_1{}^T\mathbf{e}_1} \right) \mathbf{e}_1,$$

$$\varpi = \mathbf{e}_1{}^T(f(\mathbf{x}_1) - f(\mathbf{x}_r)) + 10000\|\mathbf{e}_1\|^2.$$

The passivity degree $(\theta_{CR} + \psi_1)$ for the pinned node satisfies

$$\dot{V} = \mathbf{e}_1{}^T(f(\mathbf{x}_1) - f(\mathbf{x}_r)) - \left(\varpi + \sqrt{\varpi^2 + \left(L_g V (L_g V)^T\right)^2} \right)$$

$$= -10000\|\mathbf{e}_1\|^2 - \sqrt{\varpi^2 + \left(L_g V (L_g V)^T\right)^2}$$

$$< -10000\|\mathbf{e}_1\|^2 = -(\theta_{CR} + \psi_1)\|\mathbf{e}_1\|^2,$$

which fulfills condition (5.21), i.e., matrix $\mathbf{M}$ is negative definite.

Simulations are performed as follows. From $t = 0s$ to $t = 2.5s$, the network runs without interconnections and control actions ($c_{ij} = 0, \mathbf{u}_1 = 0$). From $t = 2.5s$, the coupling strengths are turned on, with $c_{ij} = c = 30$, without control. In this case, the network does not synchronize to the desired trajectory. Then, at $t = 5s$, the control law is applied, and the complex network gradually tracks the reference signal. Figs. 5.4(b) and 5.4(c) show the states of the entire network, $(x_{i_1}, x_{i_2}, x_{i_3}$, with $i = 1, \ldots, 9)$, with initial conditions randomly generated from the interval $[-20, 20]$. The average tracking errors $\mathbf{e}_i$ illustrate the effectiveness of the optimal pinning control scheme, showing that this controller ensures trajectory tracking for the whole complex network. Finally, Fig. 5.4(d) presents the control signals $u_{1_i}(t)$, $i = 1, 2, 3$, applied to each pinned node; they are different because each input is applied to a different state.

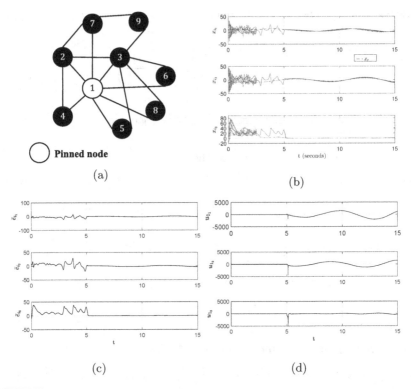

(a)

(b)

(c)   (d)

FIGURE 5.4
Results for a scale-free type of network with three types of chaotic nodes, described by the Chen system (Nodes 1, 2, and 3), Lorenz system (Nodes 4, 5, and 6), and Lü system (Nodes 7, 8, and 9): (a) Network topology; (b) Time evolution for the three states of the nine network nodes ($x_{i_1}, x_{i_2}, x_{i_3}$, with $i = 1, \ldots, 9$); (c) Average tracking error $\mathbf{e}_i$; (d) Signal of the control law $u_{1_i}(t)$, $i = 1, 2, 3$.

5.3 Conclusions

An effective controller for trajectory tracking on complex networks with chaotic nodes is presented. This pinning controller is based on an inverse optimal control law, such that the whole network tracks the desired trajectories. Additionally, this controller can manage different coupling strengths and minimizes an associated cost functional, with the main drawback that the local node systems must be known. This technique is extended by complex dynamical network with non-identical nodes. The control schemes are evaluated via simulations. The stability analysis is also presented.

Bibliography

[1] G. Chen and T. Ueta. Yet another chaotic attractor. *International Journal of Bifurcation and Chaos*, 9(7):1465–1466, 1999.

[2] M. Krstic and Z. H. Li. Inverse optimal design of input-to-state stabilizing nonlinear controllers. *IEEE Transactions on Automatic Control*, 43(3):336–350, 1998.

[3] X. Li and G. Chen. Synchronization and desynchronization of complex dynamical networks: An engineering viewpoint. *IEEE Transactions on Circuits and Systems I: Fundamental Theory and Applications*, 50(11):1381–1390, 2003.

[4] X. Li, X. Wang, and G. Chen. Pinning a complex dynamical network to its equilibrium. *IEEE Transactions on Circuits and Systems I: Regular Papers*, 51(10):2074–2087, 2004.

[5] E. N. Lorenz. Deterministic nonperiodic flow. *Journal of the Atmospheric Sciences*, 20(2):130–141, 1963.

[6] J. Lü and G. Chen. A new chaotic attractor coined. *International Journal of Bifurcation and Chaos*, 12(3):659–661, 2002.

[7] O. E. Rössler. An equation for continuous chaos. *Physics Letters A*, 57(5):397–398, 1976.

[8] E. N. Sanchez and D. I. Rodriguez. Inverse optimal pinning control for complex networks of chaotic systems. *International Journal of Bifurcation and Chaos*, 25(2):1550031, 2015.

[9] J. Xiang and G. Chen. On the V-stability of complex dynamical networks. *Automatica*, 43(6):1049–1057, 2007.

6

Neural Inverse Optimal Control

In this chapter, a new approach for trajectory tracking of uncertain complex networks with identical and non-identical nodes is introduced. To achieve this goal, a neural controller is applied to a small fraction of nodes (pinned ones). Such a controller is composed of an on-line identifier based on a recurrent high-order neural network, and an inverse optimal controller to track the desired trajectory. A complete stability analysis is also presented. In order to verify the applicability and good performance of the control scheme, representative examples are simulated, which consist of a complex network with each node described by a chaotic Lorenz oscillator, and a network with different chaotic nodes.

6.1 Trajectory Tracking of Complex Networks

This section develops a neural inverse optimal pinning control scheme for uncertain complex networks to achieve trajectory tracking. The control objective is the same as that in *Chapter 5*.

Consider a weighted network with N identical nodes, where f is replaced by $\hat{f} : \mathbb{R}^n \to \mathbb{R}^n$, and this term represents the unknown self-dynamics of node i. The control law will be a function of the identification error, the tracking

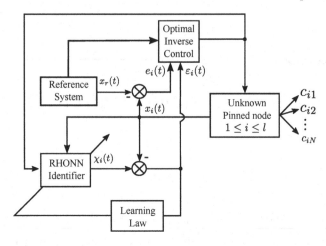

FIGURE 6.1
The neural pinning control scheme.

error and the system dynamics represented by an RHONN. This scheme is displayed in Fig. 6.1.

To achieve trajectory tracking for the whole complex network, an NN-based control law is applied to a small fraction of nodes (pinned nodes); such a controller is constituted by (a) an on-line identifier using an RHONN trained with a Gradient Algorithm as in [12, p. 20], and (b) an inverse optimal control law for trajectory tracking. The inverse optimal control law is derived on the basis of a CLF, which includes the tracking error $\mathbf{e}_i$, the identification error ε_i, and the weights estimation error of RHONN, $\tilde{W}_i$. The cost functional assigns terms to these errors and the control input. The Sontag-type control law is used as in [6, p. 339] to achieve ISS. Additionally, $\mathfrak{l}(\cdot)$ and R on the cost functional (8) are determined *a posteriori* by this control law; both of them must be positive definite and radially unbounded functions [6]. Finally, for the controlled complex network, conditions to achieve trajectory tracking are established, using the Lyapunov direct method. The requirements to satisfy all these conditions are established in Theorem 6.1.

6.1.1 Neural identifier

Consider an unknown model for the ith pinned node

$$
\begin{aligned}
\dot{\mathbf{x}}_i &= \hat{f}(\mathbf{x}_i) - \sum_{j=1}^{N} g_{ij}\mathbf{\Gamma}\mathbf{x}_j + \mathbf{u}_i \\
&\triangleq \hat{f}(\mathbf{x}_i) + \hat{h}(\mathbf{x}_i)\boldsymbol{\omega}_i + \mathbf{u}_i,
\end{aligned} \tag{6.1}
$$

where $\hat{f}(\cdot)$ and $\hat{h}(\cdot)$ are unknown smooth vector functions defined on a compact set $\mathcal{Y} \subset \mathbb{R}^n$, $\boldsymbol{\omega}_i$ are considered as disturbances, with $\mathbf{x}_i$ being available for measurement. In this section, system (6.1) is modeled by a neural network.

For identification, an RHONN in a series-parallel structure is

$$
\dot{\boldsymbol{\chi}}_i = -\lambda_i I_n \boldsymbol{\chi}_i + W_i \phi(\mathbf{x}_i) + \mathbf{u}_i, \tag{6.2}
$$

where $\lambda_i > 0$, $\boldsymbol{\chi}_i \in \mathbb{R}^n$, $\mathbf{u}_i \in \mathbb{R}^n$, and $W_i \in \mathbb{R}^{n \times L}$ represents the adjustable on-line network weights. The RHONN structure is selected, due to its capability of modeling a large class of dynamical systems. Furthermore, the number of neurons is determined by the state dynamics of the system to be identified. Additionally, this structure does not have hidden layers, therefore is very flexible and allows formal stability analysis.

Assumption 5. *For every $w_{jk} \in W_i$, (6.2) is bounded for every bounded state* $\mathbf{x}_i$.

Assumption 6. *The unknown system (6.1) can be completely described by a neural network of the form*

$$
\dot{\boldsymbol{x}}_i = -\lambda I_n \mathbf{x}_i + W_i^* \phi(\mathbf{x}_i) + \mathbf{u}_i + \boldsymbol{\omega}_{ier}, \tag{6.3}
$$

where W_i^* are the ideal optimal constant weight matrices, the modeling error $\boldsymbol{\omega}_{ier} \in \boldsymbol{\Omega} \subseteq \mathbb{R}^n$ is assumed to be bounded, and all the other elements are as described above.

The modeling error can be made arbitrarily small by selecting appropriately the number L_i of high-order connections.

Now, the identification error is defined as $\varepsilon_i = \chi_i - \mathbf{x}_i$, satisfying

$$\dot{\varepsilon}_i = -\lambda I_n \varepsilon_i + \tilde{W}_i \phi(\mathbf{x_i}) - \omega_{ier}, \qquad (6.4)$$

where $\tilde{W}_i = W_i - W_i^*$. To minimize the identification error on a compact set, a learning law to adjust the weight matrix W_i is derived, as follows [12, p. 20]:

$$tr\left\{\dot{\tilde{W}}_i^T \tilde{W}_i\right\} = -\gamma \varepsilon_i^T \tilde{W}_i \phi(\mathbf{x}_i), \qquad (6.5)$$

which has elements

$$\dot{\tilde{w}}_{ij} = -\gamma \varepsilon_i \phi(\mathbf{x}_i), \quad i = 1, 2, \ldots, n, \quad j = 1, 2, \ldots, L.$$

Following the same procedure as in [12], to analyze the stability of (6.4), consider the Lyapunov function

$$V(\varepsilon_i, \tilde{W}_i) = \frac{1}{2}\varepsilon_i^T \varepsilon_i + \frac{1}{2\gamma}tr\{\tilde{W}_i^T \tilde{W}_i\}, \quad \gamma > 0,$$

and its derivative

$$\dot{V} = \varepsilon_i^T \dot{\varepsilon}_i + \frac{1}{\gamma}tr\{\dot{\tilde{W}}_i^T \tilde{W}_i\}$$

$$= -\lambda\|\varepsilon_i\|^2 + \varepsilon_i^T(\tilde{W}_i\phi(\mathbf{x_i}) - \omega_{ier}) - \varepsilon_i^T \tilde{W}_i\phi(\mathbf{x}_i).$$

Suppose no modeling error, $\omega_{ier} = 0$, thus

$$\dot{V} \leq -\lambda\|\varepsilon_i\|^2 \leq 0$$

and applying Barbalat's lemma [4], it follows that $\lim\limits_{t\to\infty} \varepsilon_i = 0$ and $\tilde{W}_i$ is bounded. However, the assumption of no modeling error can hardly be satisfied since the higher-order terms in the RHONN model are finite. The learning law

(6.5) is modified as in [12], in order to avoid parameter drift:

$$
\dot{w}_{ij} =
\begin{cases}
-\gamma \varepsilon_i \phi(\mathbf{x}_i) & \text{if } |w_{ij}| \leq M_i, \\
-\gamma \varepsilon_i \phi(\mathbf{x}_i) - \gamma_r \gamma w_{ij} & \text{if } |w_{ij}| > M_i, \\
\end{cases}
\tag{6.6}
$$
$$
i = 1, 2, \ldots, n, \quad j = 1, 2, \ldots, L,
$$

where M_i is an upper bound for the weights and γ_r is a positive constant. With (6.6), it is guaranteed that ε_i and $\tilde{W}_i$ remain bounded, i.e.,

$$
\sup_{0 \leq t \leq T} \|\varepsilon_i(t)\| \leq \mathcal{E}
$$

for any $\mathcal{E} > 0$ and any finite $T > 0$. For a detailed stability analysis of (6.4), see [5, pp. 424–427], [12, pp. 16–24].

Once the identification error is minimized on a compact set, trajectory tracking of the neural network with modeling errors can be considered. More details about the design procedure for adaptive control based on an RHONN are included in [5, 12].

6.1.2 Control law for trajectory tracking

Suppose that unknown system (6.1) has been identified by (6.2) after a finite time T, with an associated modeling error different from zero. Due to this fact, (6.2) is used instead of (6.3), i.e., $\dot{\mathbf{x}}_i \overset{\Delta}{=} \dot{\chi}_i + \boldsymbol{\omega}_{ier}$.

$$
\dot{\mathbf{x}}_i = -\lambda I_n \chi_i + W_i \phi(\mathbf{x}_i) + \boldsymbol{\omega}_{ier} + \mathbf{u}_i
$$

and the tracking error is defined as $\mathbf{e}_i = \mathbf{x}_i - \mathbf{x}_r \in \mathbb{R}^n$. The main result is established in the following theorem, with conditions to achieve trajectory tracking via NN-based inverse optimal control for the whole network.

Theorem 6.1. *Consider the complex network with pinning control (2.14), in which $\boldsymbol{\Gamma} = \mathbf{I}_n$, and a reference system defined by (5.1). If c_{min}, the minimal* *coupling strength of the entire network, satisfies*

$$c_{min} > \frac{M}{\sigma_{min}(-A + diag(d_1(\cdot), \ldots, d_l(\cdot), 0, \ldots, 0))}, \qquad (6.7)$$

where i) $M = L_C^f > 0$ *for* $f_e(\mathbf{e}_i)$ *Lipschitz continuous in* $\mathbf{e}_i$*, or ii)* $M = h_{\max}$ *for* $f_e(\mathbf{e}_i)$ *a chaotic attractor, and*

$$d_i(\mathbf{e}_i, \tilde{W}_i) = \begin{cases} \left[\frac{\varpi + \sqrt{\varpi^2 + (\|\mathbf{e}_i\|^2)^2}}{\|\mathbf{e}_i\|^2} \right] \mathbf{e}_i, & \|\mathbf{e}_i\| \neq 0, \\ 0, & \|\mathbf{e}_i\| = 0, \end{cases}$$

with

$$\varpi = \mathbf{e}_i^T \left(W_i \phi(\mathbf{x}_i) + \alpha(\mathbf{x}_r) \right) + \|\mathbf{e}_i\| \rho^{-1} \left(\|\mathbf{e}_i\| \right), \qquad (6.8)$$

$$\alpha(\mathbf{x}_r) = \lambda I_n \mathbf{x}_r - f_r(\mathbf{x}_r),$$

where $W_i \phi(\mathbf{x}_i)$ *and* λ *are determined by the RHONN identifier (6.2), then the NN-based inverse optimal pinning control law*

$$\mathbf{u}_i = -\frac{1}{2} R^{-1}(\mathbf{e}_i, \tilde{W}_i) L_g V^T = \alpha_i^*(\mathbf{e}_i, \tilde{W}_i)$$

$$\triangleq -c_{ii} d_i(\mathbf{e}_i, \tilde{W}_i) \mathbf{e}_i, \qquad (6.9)$$

with

$$\alpha_i^* = \begin{cases} -c_{ii} d_i(\mathbf{e}_i, \tilde{W}_i), & \|\mathbf{e}_i\| \neq 0, \\ 0, & \|\mathbf{e}_i\| = 0, \end{cases}$$

guarantees that the whole complex network tracks the desired trajectory. Furthermore, (6.9) minimizes the cost functional (5.3) for all pinned nodes.

Proof. The proof is divided in three parts: the first one is the local analysis for the pinned nodes to achieve trajectory tracking; the second one is the optimality analysis; finally, the third one is the analysis of trajectory tracking for the whole network.

A. Trajectory tracking for the pinned nodes

Taking the time derivative of the tracking error $\mathbf{e}_i = \mathbf{x}_i - \mathbf{x}_r \in \mathbb{R}^n$, one has

$$
\begin{aligned}
\dot{\mathbf{e}}_i &= \dot{\mathbf{x}}_i - \dot{\mathbf{x}}_r \\
&= -\lambda I_n \chi_i + W_i \phi(\mathbf{x}_i) + \omega_{ier} + \mathbf{u}_i - f_r(\mathbf{x}_r) \\
&= -\lambda I_n \mathbf{e}_i + W_i \phi(\mathbf{x}_i) + \alpha(\mathbf{x}_r) + \boldsymbol{\omega}_{ier} + \mathbf{u}_i,
\end{aligned}
\tag{6.10}
$$

$$
\alpha(\mathbf{x}_r) = \lambda I_n \mathbf{x}_r - f_r(\mathbf{x}_r).
$$

The tracking problem is thus transformed to stabilization of error system (6.10). In order to analyze the stability, the Lyapunov method can be applied.

Consider the following Control Lyapunov Function (CLF):

$$
V(\mathbf{e}_i, \varepsilon_i, \tilde{W}_i) = \frac{1}{2}\mathbf{e}_i^T \mathbf{e}_i + \frac{1}{2}\varepsilon_i^T \varepsilon_i + \frac{1}{2\gamma} tr\{\tilde{W}_i^T \tilde{W}_i\}.
\tag{6.11}
$$

By taking its time derivative, one obtains

$$
\begin{aligned}
\dot{V} &= \mathbf{e}_i^T \dot{\mathbf{e}}_i + \varepsilon_i^T \dot{\varepsilon}_i + \frac{1}{\gamma} tr\{\dot{\tilde{W}}_i^T \tilde{W}_i\} \\
&= \mathbf{e}_i^T(-\lambda I_n \mathbf{e}_i + W_i \phi(\mathbf{x}_i) + \alpha(\mathbf{x}_r) + \omega_{ier} + \mathbf{u}_i) + \varepsilon_i^T(-\lambda I_n \varepsilon_i + \tilde{W}_i \phi(\mathbf{x}_i)) \\
&\quad - \varepsilon_i^T \tilde{W}_i \phi(\mathbf{x}_i) \\
&= -\lambda(\|\mathbf{e}_i\| + \|\varepsilon_i\|) + \mathbf{e}_i^T(W_i \phi(\mathbf{x}_i) + \alpha(\mathbf{x}_r) + \omega_{ier} + \mathbf{u}_i) \\
&\triangleq -W_1(\mathbf{e}_i, \varepsilon_i) + L_f V + L_g V \mathbf{u}_i + L_h V \boldsymbol{\delta}_i,
\end{aligned}
\tag{6.12}
$$

where

$$
\begin{aligned}
W_1(\mathbf{e}_i, \varepsilon_i) &= \lambda(\|\mathbf{e}_i\| + \|\varepsilon_i\|) > 0, \\
L_f V &= \mathbf{e}_i^T \left(W_i \phi(\mathbf{x}_i) + \alpha(\mathbf{x}_r) \right), \\
L_g V &= \mathbf{e}_i^T, \\
L_h V &= \mathbf{e}_i^T, \quad \text{with} \quad \boldsymbol{\delta}_i = \omega_{ier}.
\end{aligned}
$$

Next, the following control law $\mathbf{u}_i$, taken from [6], is used:

$$\mathbf{u}_i = -\frac{1}{2}R^{-1}(\mathbf{e}_i, \tilde{W}_i)L_g V^T = \alpha_i^*(\mathbf{e}_i, \tilde{W}_i)$$
$$\triangleq -c_{ii}d_i(\mathbf{e}_i, \tilde{W}_i)\mathbf{e}_i, \tag{6.13}$$

where

$$\alpha_i^* = \begin{cases} -c_{ii}\left[\dfrac{\varpi + \sqrt{\varpi^2 + \left(L_g V(L_g V)^T\right)^2}}{L_g V(L_g V)^T}\right] L_g V^T, & \|L_g V\| \neq 0, \\[4mm] 0, & \|L_g V\| = 0, \end{cases}$$

with

$$\varpi = L_f V + |L_h V|\, \rho^{-1}\left(\|\mathbf{e}_i\|\right), \tag{6.14}$$

and

$$d_i(\mathbf{e}_i, \tilde{W}_i) = \frac{\varpi + \sqrt{\varpi^2 + \left(L_g V(L_g V)^T\right)^2}}{L_g V(L_g V)^T}.$$

Replacing (6.9) in (6.12) gives

$$\dot{V}\Big|_{u=\frac{\alpha_i^*}{2}} = -W_1(\mathbf{e}_i, \varepsilon_i) + L_f V + L_h V \delta_i$$
$$- \frac{c_{ii}}{2}\left(\varpi + \sqrt{\varpi^2 + \left(L_g V(L_g V)^T\right)^2}\right)$$
$$= -W_1(\mathbf{e}_i, \varepsilon_i) + L_h V \delta_i - |L_h V|\, \rho^{-1}\left(\|\mathbf{e}_i\|\right)$$
$$- \frac{c_{ii}}{2}\left(\sqrt{\varpi^2 + \left(L_g V(L_g V)^T\right)^2}\right)$$
$$+ \left(1 - \frac{c_{ii}}{2}\right)\overbrace{\left(L_f V + |L_h V|\, \rho^{-1}\left(\|\mathbf{e}_i\|\right)\right)}^{\varpi}$$
$$\leq -W_1(\mathbf{e}_i, \varepsilon_i) - |L_h V|\left(\rho^{-1}\left(\|\mathbf{e}_i\|\right) - \|\delta_i\|\right) + \left(1 - \frac{c_{ii}}{2}\right)\varpi$$
$$- \frac{c_{ii}}{2}\sqrt{\varpi^2 + \frac{1}{2}\left(L_g V(L_g V)^T\right)^2}.$$

For $\|\mathbf{e}_i\| \geq \rho(\|\boldsymbol{\delta}_i\|)$, one has

$$\dot{V}(\mathbf{e}_i, \varepsilon_i, \tilde{W}_i) \leq -W_1(\mathbf{e}_i, \varepsilon_i) - W_2(\mathbf{e}_i, \tilde{W}_i) < 0,$$

where

$$W_2(\mathbf{e}_i, \tilde{W}_i) = \left(\frac{c_{ii}}{2} - 1\right)\varpi + \frac{c_{ii}}{2}\sqrt{\varpi^2 + \frac{1}{2}\left(L_g V (L_g V)^T\right)^2},$$

which is positive definite for all $c_{ii} > 1$. Hence, error system (6.10) is input-to-state stable (ISS) [6].

B. Inverse optimality

Once stability is guaranteed, using control law (6.9), inverse optimality can be analyzed. It is required that this control law minimizes the cost functional (5.3). Consequently, Lyapunov function (6.11) must satisfy the associated HJI equation:

$$\mathfrak{l}(\mathbf{e}_i, \varepsilon_i, \tilde{W}_i) - W_1(\mathbf{e}_i, \varepsilon_i) + L_f V - \frac{1}{4}L_g V R^{-1}(L_g V)^T + \|L_h V\|\boldsymbol{\delta}_i = 0. \quad (6.15)$$

When $\mathfrak{l}(\mathbf{e}_i, \varepsilon_i, \tilde{W}_i)$ is selected as $\mathfrak{l}(\mathbf{e}_i, \varepsilon_i, \tilde{W}_i) = -\dot{V}$, function (6.11) is a solution to (6.15). Hence, $\mathfrak{l}(\mathbf{e}_i, \varepsilon_i, \tilde{W}_i)$ is written as

$$\mathfrak{l}(\mathbf{e}_i, \varepsilon_i, \tilde{W}_i) = W_1(\mathbf{e}_i, \varepsilon_i) - L_f V + \frac{1}{4}L_g V R^{-1}(L_g V)^T - \|L_h V\|\boldsymbol{\delta}_i,$$

where $\mathfrak{l}(\mathbf{e}_i, \varepsilon_i, \tilde{W}_i)$ must be positive definite and radially unbounded. Substituting $R^{-1}(\mathbf{e}_i, \tilde{W}_i)$ into (6.16) yields

$$\mathfrak{l}(\mathbf{e}_i, \varepsilon_i, \tilde{W}_i) \geq W_1(\mathbf{e}_i, \varepsilon_i) + W_2(\mathbf{e}_i, \tilde{W}_i) > 0,$$

for $\|\mathbf{e}_i\| \geq \rho(\|\boldsymbol{\delta}_i\|)$.

Clearly, $\mathfrak{l}(\mathbf{e}_i, \varepsilon_i, \tilde{W}_i)$ is positive definite and radially unbounded, guarantying that control law (6.9) minimizes cost functional (5.3).

Substituting (6.16) and $\mathbf{u}_i = \mathbf{v}_i - \frac{1}{2}R^{-1}(\mathbf{e}_i, \tilde{W}_i)L_g V^T$ into (5.3) gives

$$
\begin{aligned}
J_i &= \lim_{t\to\infty} \left[V + \int_0^t \left(l(\mathbf{e}_i, \boldsymbol{\varepsilon}_i, \tilde{W}_i) + \mathbf{v}_i^T R \mathbf{v}_i - \mathbf{v}_i^T (L_g V)^T + \frac{1}{4} L_g V R^{-1} (L_g V)^T \right) d\tau \right] \\
&= \lim_{t\to\infty} \left[V + \int_0^t \left(W_1(\mathbf{e}_i, \boldsymbol{\varepsilon}_i) - L_f V + \frac{1}{2} L_g V R^{-1} (L_g V)^T \right. \right. \\
&\qquad\qquad \left. \left. -(L_g V)\mathbf{v}_i - \|L_h V\|\boldsymbol{\delta}_i \right) d\tau + \int_0^t \left(\mathbf{v}_i^T R \mathbf{v}_i \right) d\tau \right] \\
&= \lim_{t\to\infty} \left[V - \int_0^t \left(\sup_{\boldsymbol{\delta}_i \in \Delta} \left\{ \dot{V}(\mathbf{e}_i, \boldsymbol{\varepsilon}_i, \tilde{W}_i) d\tau \right\} \right) + \int_0^t \left(\mathbf{v}_i^T R \mathbf{v}_i \right) d\tau \right] \\
&\leq \lim_{t\to\infty} \left[V - \int_0^t \left(\dot{V}(\mathbf{e}_i, \boldsymbol{\varepsilon}_i, \tilde{W}_i) \right) d\tau + \int_0^t \left(\mathbf{v}_i^T R \mathbf{v}_i \right) d\tau \right] \\
&\leq V(\mathbf{e}_i(0), \boldsymbol{\varepsilon}_i(0), \tilde{W}_i(0)) + \lim_{t\to\infty} \left[\int_0^t \left(\mathbf{v}_i^T R \mathbf{v}_i \right) d\tau \right].
\end{aligned}
$$

The minimum value of $J_i = J_i^*$ is achieved when $\mathbf{v}_i \equiv 0$, giving

$$
J_i^* \;\leq\; V(\mathbf{e}_i(0), \boldsymbol{\varepsilon}_i(0), \tilde{W}_i(0)). \tag{6.16}
$$

To complete the optimality analysis, it is necessary to ensure that $J_i^* = V(\mathbf{e}_i(0), \boldsymbol{\varepsilon}_i(0), \tilde{W}_i(0))$. For this purpose, considering an admissible disturbance $\boldsymbol{\delta}_\theta \in \Delta$, where for all $\mathbf{e}_{i_\theta}(0), \boldsymbol{\varepsilon}_{i_\theta}(0), \tilde{W}_{i_\theta}(0), \mathbf{u}_i, \theta > 0$, one has

$$
\int_0^t \dot{V}(\mathbf{e}_{i_\theta}, \boldsymbol{\varepsilon}_{i_\theta}, \tilde{W}_{i_\theta}) d\tau \geq \int_0^t \sup_{\boldsymbol{\delta}_i \in \Delta} \dot{V}(\mathbf{e}_{i_\theta}, \boldsymbol{\varepsilon}_{i_\theta}, \tilde{W}_{i_\theta}) d\tau - \theta.
$$

It follows that

$$
\begin{aligned}
J_i^* &= \lim_{t\to\infty} \left[V - \int_0^t \sup_{\boldsymbol{\delta}_i \in \Delta} \dot{V}(\mathbf{e}_{i_\theta}, \boldsymbol{\varepsilon}_{i_\theta}, \tilde{W}_{i_\theta}) d\tau \right] \\
&\geq \lim_{t\to\infty} \left[V - \int_0^t \dot{V}(\mathbf{e}_{i_\theta}, \boldsymbol{\varepsilon}_{i_\theta}, \tilde{W}_{i_\theta}) d\tau \right] - \theta \\
&\geq V(\mathbf{e}_i(0), \boldsymbol{\varepsilon}_i(0), \tilde{W}_i(0)) - \theta. \tag{6.17}
\end{aligned}
$$

Finally, (6.16) and (6.17) together results in

$$V(\mathbf{e}_i(0), \varepsilon_i(0), \tilde{W}_i(0)) - \theta \le J_i^* \le V(\mathbf{e}_i(0), \varepsilon_i(0), \tilde{W}_i(0)).$$

For an arbitrarily small θ, one has $J_i^* = V(\mathbf{e}_i(0), \varepsilon_i(0), \tilde{W}_i(0))$. Hence, control law (6.9) is optimal with respect to the cost functional (5.3).

C. Trajectory tracking for the whole network

This analysis proceeds similarly to [8].

Replacing (6.9) in (2.14), and supposing that $\mathbf{\Gamma} = I_n$, the dynamics for the tracking error $\dot{\mathbf{e}}_i = \dot{\mathbf{x}}_i - \dot{\mathbf{x}}_r$ of whole network can be written as

$$\dot{\mathbf{e}}_i = \hat{f}(\mathbf{x}_i) - f_r(\mathbf{x}_r) + \sum_{j=1}^{N} c_{ij}a_{ij}\mathbf{e}_j - b_ic_{ii}d_i(\mathbf{e}_i, \tilde{W}_i)\mathbf{e}_i,$$
$$i = 1, 2, \ldots, N.$$

Define
$$D' = diag\,(c_{11}d_1(\cdot), \ldots, c_{ll}d_l(\cdot), 0, \cdots, 0) \in \mathbb{R}^{N \times N},$$
$$G = (g_{ij}) = -(c_{ij}a_{ij}) \in \mathbb{R}^{N \times N}.$$

Furthermore, using the Kronecker product, write

$$\dot{\mathbf{E}} = I_N \otimes f_e(\mathbf{e}_i) - [(G + D') \otimes I_n]\,\mathbf{E}, \tag{6.18}$$

where $\mathbf{E} = [\mathbf{e}_1^T, \ldots, \mathbf{e}_N^T]^T \in \mathbb{R}^{Nn}$, $f_e(\mathbf{e}_i) = \hat{f}(\mathbf{e}_i - \mathbf{x}_r) - f(\mathbf{x}_r) \in \mathbb{R}^n$, G is a symmetric and semi-positive definite matrix, and $G + D'$ is positive definite with the minimal eigenvalue $\sigma_{min}(G+D') > 0$. For stability analysis of (6.18), two cases are considered.

Case 1: Assume that $f_e(\mathbf{e}_i)$ is Lipschitz continuous in $\mathbf{e}_i$ with a constant $L_C^f > 0$, and let H be a square matrix such that $f_e(\mathbf{e}_i) + H\mathbf{e}_i$ is V-uniformly decreasing for some symmetrical and positive definite matrix V and for all $\mathbf{e}_i \in \mathbb{R}^n$, $i = 1, 2, \ldots, N$. It can be shown that if there exists a positive-definite diagonal matrix $U \in \mathbb{R}^{N \times N}$ such that

$$(U \otimes V)\left[(G + D') \otimes I_n + I_N \otimes H\right] > 0, \qquad (6.19)$$

then system (6.18) is ISS about $\mathbf{e}_i = 0$.

To verify this, a Lyapunov function is chosen as

$$V_N = \frac{1}{2}\mathbf{E}^T(U \otimes V)\mathbf{E}. \qquad (6.20)$$

By taking its time derivative, one obtains

$$
\begin{aligned}
\dot{V}_N =&\, \mathbf{E}^T(I_N \otimes V)\dot{\mathbf{E}} \\
=&\, \mathbf{E}^T(U \otimes V)(I_N \otimes f_e(\mathbf{e}_i) - \left[(G + D') \otimes I_n\right]\mathbf{E}) \\
=&\, \mathbf{E}^T(U \otimes V)(I_N \otimes f_e(\mathbf{e}_i) + (I_N \otimes H)\mathbf{E}) \\
&\, - \mathbf{E}^T(U \otimes V)\left[(G + D') \otimes I_n + I_N \otimes H\right]\mathbf{E} \\
\leq&\, \mathbf{E}^T(U \otimes V)(I_N \otimes f_e(\mathbf{e}_i) + (I_N \otimes H)\mathbf{E}) \\
\leq&\, \sum_{j=1}^{N} u_j \mathbf{e}_j^T V[f_e(\mathbf{e}_j) - \alpha_e I_n \mathbf{e}_j], \quad H = -\alpha_e I_n \\
\leq&\, -\sum_{j=1}^{N} u_j \sigma_{min}(V)[\alpha_e - L_C^f]\|\mathbf{e}_j\|^2, \quad \alpha_e > L_C^f \\
\leq&\, -\sum_{j=1}^{N} \alpha_E \|\mathbf{e}_j\|^2 < 0, \quad \alpha_E > 0.
\end{aligned}
$$

For (6.19) to be positive definite, the following inequality must be satisfied:

$$\sigma_{min}(G + D') > \alpha_e > L_C^f, \qquad (6.21)$$

thus the following inequality holds:

$$
\begin{aligned}
&\sigma_{min}\left(c_{min}\left[-A + diag(d_1(\cdot), \ldots, d_l(\cdot), 0, \ldots, 0)\right]\right) \\
&\leq \sigma_{min}\left(G + diag(c_{min}d_1(\cdot), \ldots, c_{min}d_l(\cdot), 0, \ldots, 0)\right). \qquad (6.22)
\end{aligned}
$$

Based on the fact that $c_{ij} \geq c_{min} > 0$ for all c_{ij} in G, both (6.21) and (6.22) hold, so that

$$c_{min} > \frac{L_C^f}{\sigma_{min}\left([-A + diag(d_1(\cdot), \ldots, d_l(\cdot), 0, \ldots, 0)]\right)}. \tag{6.23}$$

Hence, system (6.18) is ISS at $\mathbf{e}_i = 0$.

Case 2: Assume that $f_e(\mathbf{e}_i)$ is chaotic for all $i = 1, 2, \ldots, N$, with the maximum positive Lyapunov exponent $h_{max} > 0$. The condition is that c_{min}, the minimal coupling strength of the entire network, satisfies

$$c_{min} > \frac{h_{max}}{\sigma_{min}\left([-A + diag(d_1(\cdot), \ldots, d_l(\cdot), 0, \ldots, 0)]\right)}. \tag{6.24}$$

To verify this criterion, network (6.18) is linearized in an outer neighborhood of the ultimate bound, so that

$$\dot{\mathbf{E}} = \mathbf{E}[Df_e(0)] - \hat{B}\mathbf{E},$$

where $Df_e(0) \in \mathbb{R}^{n \times n}$ is the Jacobian of $f_e(\mathbf{e}_i)$ with $\hat{B} = G + D'$. Note that $\hat{B}$ is symmetric and positive definite; hence, all of its eigenvalues are positive.

According to [14], one has

$$\sigma_{min}\left(c_{min}\left[-A + diag(d_1(\cdot), \ldots, d_l(\cdot), 0, \ldots, 0)\right]\right)$$
$$\leq \sigma_{min}\left(G + diag(c_{min}d_1(\cdot), \ldots, c_{min}d_l(\cdot), 0, \ldots, 0)\right).$$

This inequality is obtained based on the fact that $c_{ij} \geq c_{min} > 0$ for all c_{ij} in G.

Furthermore, the Transversal Lyapunov Exponents (TLEs), for all eigenvalues σ_i, $i = 1, 2, \ldots, N$, are given by $\mu_k(\sigma_i) = h_k - c\sigma_i$, $k = 1, 2, \ldots n$, which determine the stability of the controlled states [7]. Hence, local stability of network (2.14) ensures negative TLEs. Thus, the following condition must be satisfied:

$$\mu_{\max}(\sigma_{\min}) = h_{\max} - c_{\min}\sigma_{min}(-A + diag(d_1(\cdot), \ldots, d_l(\cdot), 0, \ldots, 0)) < 0.$$

Consequently, with the control law (6.9), trajectory tracking is achieved by the whole complex network.   □

6.1.3   Simulation results

To verify the control scheme for its applicability, an illustrative example of a scale-free Barabási–Albert network constituting of 40 nodes is simulated. In this network, each node is selected as the Lorenz system, which is a three-dimensional autonomous chaotic system with complex dynamical behaviors, given by [9]. Fig. 6.2(a) presents the phase portrait of the Lorenz oscillator for different parameters: 1. ($a_L = 10$, $b_L = 28$, and $c_L = 8/3$), 2. ($a_L = 10$, $b_L = 56$, and $c_L = 8/3$), and 3. ($a_L = 10$, $b_L = 84$, and $c_L = 8/3$).

This example is selected as a common practice to illustrate the capabilities of the control scheme. Additionally, it presents chaotic dynamics on each node, which is complex enough for testing a new control scheme in a general study.

Let $\boldsymbol{\Theta} = diag(1,1,1)$, and define the state variables as: $x_1 = x$, $x_2 = y$, $x_3 = z$. The equation of the pinned node $\mathbf{x}_1$ is given by

$$\dot{x}_{1_1} = a_L(x_{1_2} - x_{1_1}) + \sum_{j=1}^{40} c_{1j}a_{1j}x_{j_1} + u_{1_1},$$

$$\dot{x}_{1_2} = x_{1_1}(b_L - x_{1_3}) - x_{1_2} + \sum_{j=1}^{40} c_{1j}a_{1j}x_{j_2} + u_{1_2}, \qquad (6.25)$$

$$\dot{x}_{1_3} = x_{1_1}x_{1_2} - c_L x_{1_3} + \sum_{j=1}^{40} c_{1j}a_{1j}x_{j_3} + u_{1_3}.$$

Simulations are performed as follows. From $t = 0s$ to $t = 2.3s$, the network runs without any connection ($c_{ij} = 0$). At $t = 2.5s$, the coupling strengths are selected randomly with $c_{ij} > 50$. Then, at $t = 2.5s$, the controller is applied, and a reference trajectory is selected as the output of the Lü system [10]. At $t = 5s$, plant parameter changes are imposed for the network odd-numbered nodes as follows:

$$b_L = \begin{cases} 28, & 0 < t \leq 5, \text{ and } 8 < t \leq 9 \\ 56, & 5 < t \leq 6, \ 7 < t \leq 8, \text{ and } 9 < t \leq 10 \\ 84, & 6 < t \leq 7 \end{cases}$$

The controller is defined as in (6.9) and applied to Node 1, which is the node with the highest degree. For simulations, the following RHONN is used to model the pinned node:

$$\dot{\chi}_{1_1} = -50\chi_{1_1} + W_1 z_1(\mathbf{x}_1) + u_{1_1},$$
$$\dot{\chi}_{1_2} = -50\chi_{1_2} + W_2 z_2(\mathbf{x}_1) + u_{1_2},$$
$$\dot{\chi}_{1_3} = -50\chi_{1_3} + W_3 z_3(\mathbf{x}_1) + u_{1_3}$$

with

$$W_1 = [w_{11}, w_{12}, w_{13}],$$
$$W_2 = [w_{21}, w_{22}, w_{23}, w_{24}],$$
$$W_3 = [w_{31}, w_{32}, w_{33}],$$

and

$$z_1(\mathbf{x}_i) = [tanh(x_{1_1}); tanh(x_{1_2}); tanh^3(x_{1_1})],$$
$$z_2(\mathbf{x}_i) = [tanh(x_{1_1}); tanh(x_{1_2}); tanh(x_{1_1})tanh(x_{1_3}); tanh^3(x_{1_2})],$$
$$z_3(\mathbf{x}_i) = [tanh(x_{1_3}); tanh(x_{1_1})tanh(x_{1_2}); tanh^3(x_{1_3})].$$

Figs. 6.2(b) and 6.2(c) exhibit the evolutions of the network states and their average, respectively. Fig. 6.2(d) presents the values of the coupling strengths $c_{ii} = \frac{1}{k_i} \sum\limits_{j=1, j \neq i}^{N} c_{ij} a_{ij}$. The time-varying effects of the interconnected nodes and the coupling strengths can be seen as unmodeled dynamics and external disturbances. Fig. 6.3(a) presents the control input signal $\mathbf{u}_i(t)$ applied to the pinned nodes. The tracking errors $\mathbf{e}_i$ of the pinned node are displayed in Fig. 6.3(b). Figs. 6.3(c) and 6.3(d) display the evolutions of identification errors ε_i and weights for Node 1, respectively, which are bounded. These results illustrate good identifier performance.

As can be seen in this example, tracking is achieved even in the presence of plant parameter changes, unmodeled dynamics, and bounded external disturbances, illustrating the robustness of the controller.

Table 6.1 presents statistical measures (Mean Square Error, Standard Deviation, and Maximum Value) of the NN-based inverse optimal pinning controller (the identification error ε_i, the tracking error for the pinned node (Node 1) $\mathbf{e}_i$, and the average tracking error $\bar{\mathbf{e}}_i$).

TABLE 6.1

Results of statistical measures of the NN-based inverse optimal pinning controller.

Measure	ε_i	$\mathbf{e}_i$	$\bar{\mathbf{e}}_i$
Mean Square Error	1.2878	4.1494×10^{-4}	0.7677
Standard Deviation	1.1348	0.0204	0.8739
Maximum Value	8.3690	5.8001	3.7589

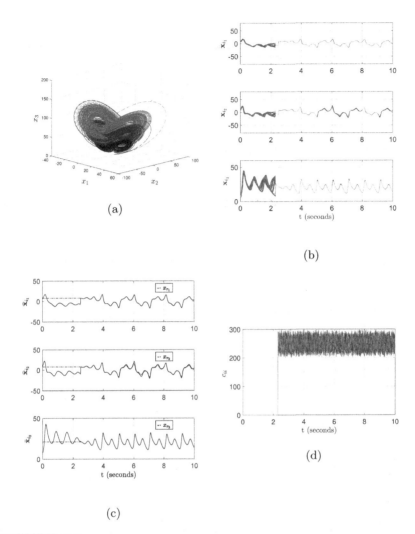

FIGURE 6.2

Results for a scale-free constituting of 40 nodes, where each network node selected as the chaotic Lorenz system: (a) Phase portrait of Lorenz's oscillator for different parameters: 1. ($a_L = 10$, $b_L = 28$, and $c_L = 8/3$), 2. ($a_L = 10$, $b_L = 56$, and $c_L = 8/3$), and 3. ($a_L = 10$, $b_L = 84$, and $c_L = 8/3$); (b) The evolution of network states; (c) Average trajectory of network states, $\bar{\mathbf{x}}_i$ (- continuous line), reference trajectory, $\mathbf{x}_r$ (−· dashed line); (d) Coupling strengths c_{ii}, $i = 1, 2, \ldots, 40$.

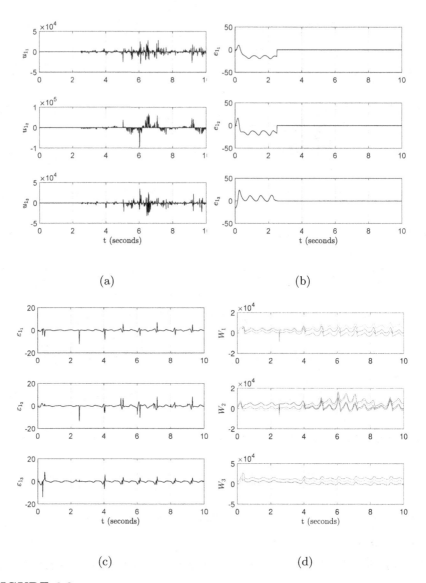

(a)   (b)

(c)   (d)

FIGURE 6.3

Results for performance of the NN-based inverse optimal pinning control law: (a) Signal of the control law $u_{1_i}(t)$, $i = 1, 2, 3$; (b) Tracking error $\mathbf{e}_1$; (c) Identification error $\boldsymbol{\varepsilon}_1$; (d) Weights evolution at Node 1 with identification.

6.2 Non-Identical Nodes

The following Lemma addresses the neural inverse optimal pinning control for trajectory tracking in complex networks with nonidentical nodes.

Lemma 6.1. *The complex network* (2.21) *tracks the desired trajectory* (5.1) *if the condition*

$$\mathbf{M} = (-\mathbf{\Theta} + \mathbf{G} - \mathbf{\Psi}) \otimes \mathbf{P\Gamma} < 0 \tag{6.26}$$

is satisfied, where $\mathbf{\Theta} = diag(\theta_1, \theta_2, \ldots, \theta_N) \in \mathbb{R}^{N \times N}$, $\mathbf{G} = (g_{ij}) = (c_{ij}a_{ij}) \in \mathbb{R}^{N \times N}$, $\mathbf{\Psi} \in \mathbb{R}^{N \times N}$ *is a diagonal matrix with the first* l *elements* ψ_i, $i = 1, 2, \ldots, l$, *and the other* $(N - l)$ *elements are all zero,* $\mathbf{P} = \mathbf{P}^T > 0 \in \mathbb{R}^{n \times n}$, *with NN-based inverse optimal pinning control law* (6.9). *Furthermore,* (6.9) *minimizes the cost functional* (5.3) *for all pinned nodes.*

Proof. It is completed along the same line as the procedure used for Lemma 5.1; so, it is omitted. □

To verify the control applicability, consider a network model composing of 6 nodes with three types of nodes, which are described by the Chen system (Nodes 1 and 2) [3], Lorenz system (Nodes 3 and 4) [9], and Lü system (Nodes 5 and 6) [10]. The coupling strengths are constants $c_{ij} = c = 30$, and $\mathbf{\Gamma} = diag(1, 1, 1)$. The control goal is to track a reference signal from the Rössler system (2.12)

The network matrix $\mathbf{A}$ is

$$A = \begin{bmatrix} -5 & 1 & 1 & 1 & 1 & 1 \\ 1 & -3 & 1 & 1 & 0 & 0 \\ 1 & 1 & -4 & 0 & 1 & 1 \\ 1 & 1 & 0 & -2 & 0 & 0 \\ 1 & 0 & 1 & 0 & -2 & 0 \\ 1 & 0 & 1 & 0 & 0 & -2 \end{bmatrix}.$$

Only one controller is necessary to stabilize the network to the equilibrium point $\bar{\mathbf{x}}$. This controller is applied to Node 1. For simulations, the following RHONN is used to model the pinned node:

$$
\begin{aligned}
\dot{\chi}_{i_1} &= -2\chi_{i_1} + W_1 z_1(\mathbf{x}_i) + u_{i_1} \\
\dot{\chi}_{i_2} &= -2\chi_{i_2} + W_2 z_2(\mathbf{x}_i) + u_{i_2} \\
\dot{\chi}_{i_3} &= -2\chi_{i_3} + W_3 z_3(\mathbf{x}_i) + u_{i_3}
\end{aligned}
$$

with
$$
\begin{aligned}
z_1(\mathbf{x}_i) &= [tanh(x_{i_1}); tanh(x_{i_2}); tanh^2(x_{i_1})], \\
z_2(\mathbf{x}_i) &= [tanh(x_{i_1}); tanh(x_{i_2}); tanh(x_{i_1})tanh(x_{i_3})], \\
z_3(\mathbf{x}_i) &= [tanh(x_{i_3}); tanh(x_{i_1})tanh(x_{i_2})].
\end{aligned}
$$

Simulations are performed as follows. From $t = 0s$ to $t = 1s$, the network runs without interconnections and control actions ($c_{ij} = 0, \mathbf{u}_1 = 0$). From $t = 1s$, the coupling strengths are selected as $c_{ij} = c = 30$, without control and it does not synchronize to the desired point. Then, at $t = 2s$, the control law is turned on, and the complex network tracks the desired trajectory.

Figs. 6.4(a) and 6.4(b) exhibit the evolutions of the network states and their average, respectively. Fig. 6.5(a) presents the control input signal $\mathbf{u}_i(t)$ applied to the pinned nodes. The tracking errors $\mathbf{e}_i$ of the pinned node are displayed in 6.5(b). Figs. 6.5(c) and 6.5(d) display the evolutions of identification errors ε_i and weights for Node 1, respectively, which are bounded. These results illustrate good identifier performance.

6.3   Discrete-Time Case

In this section, a sufficient condition for trajectory tracking is derived to ensure the stability of the error system, using a sampled-data controller.

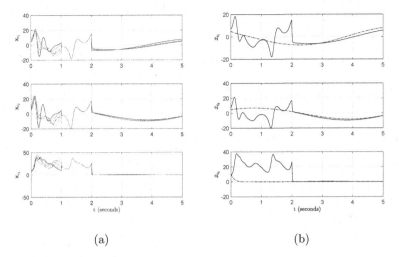

(a)   (b)

FIGURE 6.4
Results of scale-free network model with three types of nodes, which are described by Chen's system (Nodes 1 and 2) [3], Lorenz's system (Nodes 3 and 4) [9], and Lü's system (Nodes 5 and 6) [10]: (a) The evolution of network states; (b) Average trajectory of network states, $\bar{\mathbf{x}}_i$ (- continuous line), reference trajectory, $\mathbf{x}_r$ ($-\cdot$ dashed line)

Consider a control signal $\mathbf{u}_i$, for the continuous system, generated by a zero-order hold function

$$\mathbf{u}_i(t) = \mathbf{u}_{d_i}(t_k), \quad t_k \le t \le t_k + 1$$

with a sequence of hold times

$$0 = t_0 < t_1 < \cdots < t_k < \cdots, \quad \lim_{k \to \infty} t_k = \infty,$$

where $\mathbf{u}_{d_i}$ is a discrete-time control signal. It is assumed that $t_{k+1} - t_k \le h_k \le h$, where $h > 0$. A control law of the following form is adopted:

$$\mathbf{u}_i(t) = \mathbf{u}_{d_i}(t_k) = h_i(\mathbf{e}_i(t), \mathbf{e}_i(t - \tau(t))), \quad \tau(t) = t - t_k, \tag{6.27}$$

where $h_i : \mathbb{R}^n \times \mathbb{R}^n \to \mathbb{R}^n$, and the delay $\tau(t) \le 0$ is piecewise-continuous with

$\dot{\tau} = 1$ for $t \neq t_k$. The controlled network (2.21), with (6.27) becomes

$$\dot{\mathbf{e}}_i = f_i(\mathbf{e}_i + \mathbf{x}_r) - f_r(\mathbf{x}_r) + \sum_{j=1}^{N} c_{ij} a_{ij} \boldsymbol{\Gamma} \mathbf{e}_j + b_{ii} h_i(\mathbf{e}_i(t), \mathbf{e}_i(t - \tau(t))), \quad (6.28)$$

$$\tau(t) = t - t_k, \quad t \in [t_k, t_k + 1), \quad i = 1, 2, \ldots, N.$$

The objective is to find a Lyapunov functional $V : D \subseteq \mathbb{R} \times [-h, 0] \mapsto \mathbb{R}_+$, such that for each node function $f_i(\cdot)$, there is a scalar θ_i satisfying

$$\frac{\partial V(\mathbf{e}_i)}{\partial \mathbf{e}_i} \Big[f_i(\mathbf{e}_i + \mathbf{x}_r) - f_r(\mathbf{x}_r) + b_{ii} h_i(\mathbf{e}_i(t), \mathbf{e}_i(t - \tau(t)))$$

$$+ (\theta_i + b_{ii} \psi_i) \boldsymbol{\Gamma} \mathbf{e}_i \Big] < 0,$$

$$\forall \mathbf{e}_i \in D_i, \quad \mathbf{e}_i \neq 0, \quad (6.29)$$

with constants $\psi_i \geq 0$, where θ_i represents the passivity degree and

$$D_i = \big\{ \mathbf{e}_i : \|\mathbf{e}_i\| < \delta \big\}, \quad \delta > 0, \quad D = \bigcup_{i=1}^{N} D_i.$$

Theorem 6.2. *The controlled network (2.21) tracks the desired reference defined by (5.1) if the following LMI (6.30) holds:*

$$\mathbf{M} = \begin{bmatrix} (\mathbf{G} - \boldsymbol{\Theta} - \boldsymbol{\Psi}) \otimes \mathbf{P} \boldsymbol{\Gamma} + I_N \otimes \mathbf{Q} & 0 \\ 0 & -(1 - d) I_N \otimes \mathbf{Q} \end{bmatrix} < 0, \quad (6.30)$$

where $\boldsymbol{\Theta} = diag(\theta_1, \theta_2, \ldots, \theta_N) \in \mathbb{R}^{N \times N}$, $\mathbf{G} = (g_{ij}) = (c_{ij} a_{ij}) \in \mathbb{R}^{N \times N}$, $\boldsymbol{\Psi} \in \mathbb{R}^{N \times N}$ *is a diagonal matrix with the first l elements* $\psi_i \neq 0$, $i = 1, 2, \ldots, l$, *and all the other* $(N - l)$ *elements are equal to zero,* $\mathbf{P} = \mathbf{P}^T > 0 \in \mathbb{R}^{n \times n}$, $\mathbf{Q} > 0 \in \mathbb{R}^{n \times n}$, *and the control law is given by (6.27).*

Proof. A Lyapunov function for the error system (6.28) is constructed as

$$V_N(t) = \sum_{i=1}^{N} V_i(t)$$

$$= \sum_{i=1}^{N} \left[\frac{1}{2} \mathbf{e}_i(t)^T \mathbf{P} \mathbf{e}_i(t) + \int_{t-\tau(t)}^{t} \mathbf{e}_i^T(s) \mathbf{Q} \mathbf{e}_i(s) ds \right],$$

$$t \in [t_k, t_k + 1).$$

Differentiating V_N along (6.28) gives

$$\dot{V}_N(t) = \sum_{i=1}^{N} \frac{\partial V_i(\mathbf{e}_i)}{\partial \mathbf{e}_i} \Big[f_i(\mathbf{e}_i(t) + \mathbf{x}_r(t)) - f_r(\mathbf{x}_r(t))$$

$$+ \sum_{j=1}^{N} c_{ij} a_{ij} \mathbf{\Gamma} \mathbf{e}_j(t) + b_{ii} g_i(\mathbf{e}_i(t), \mathbf{e}_i(t - \tau(t))) \Big].$$

Equation (6.29) implies that, for $\mathbf{E} = [\mathbf{e}_1^T, \mathbf{e}_2^T, \dots, \mathbf{e}_N^T]^T$,

$$\dot{V}_N(t) < \sum_{i=1}^{N} \frac{\partial V_i(\mathbf{e}_i)}{\partial \mathbf{e}_i} \left(\sum_{j=1}^{N} c_{ij} a_{ij} \mathbf{\Gamma} \mathbf{e}_j(t) - (\theta_i + b_{ii} \psi_i) \mathbf{\Gamma} \mathbf{e}_i(t) \right)$$

$$< \sum_{i=1}^{N} \mathbf{e}_i^T(t) \mathbf{P} \left(\sum_{j=1}^{N} c_{ij} a_{ij} \mathbf{\Gamma} \mathbf{e}_j(t) - (\theta_i + b_{ii} \psi_i) \mathbf{\Gamma} \mathbf{e}_i(t) \right)$$

$$+ \sum_{i=1}^{N} \Big[\mathbf{e}_i^T(t) \mathbf{Q} \mathbf{e}_i(t) - (1-d) \mathbf{e}_i^T(t - \tau) \mathbf{Q} \mathbf{e}_i(t - \tau) \Big]$$

$$< \mathbf{E}^T(t) (\mathbf{G} - \mathbf{\Theta} - \mathbf{\Psi}) \otimes \mathbf{P} \mathbf{\Gamma} \mathbf{E}(t) + \mathbf{E}^T(t) I_N \otimes \mathbf{Q} \mathbf{E}(t)$$

$$- (1-d) \mathbf{E}^T(t - \tau) I_N \otimes \mathbf{Q} \mathbf{E}(t - \tau)$$

$$< [\mathbf{E}^T(t) \ \mathbf{E}^T(t - \tau)] \mathbf{M} \begin{bmatrix} \mathbf{E}(t) \\ \mathbf{E}(t - \tau) \end{bmatrix}$$

$$\le - \alpha \| \mathbf{E}(t) \|^2.$$

If (6.30) holds for all $\alpha > 0$, then the stability of the error system (6.28) is ensured   □

A discrete-time controller is designed to achieve the control objective (trajectory tracking). This controller is composed of a discrete-time on-line neural identifier based on an RHONN and a discrete-time inverse optimal controller for each pinned node. This scheme is displayed in Fig. 6.6.

6.3.1   Discrete-time on-line neural identifier

Consider an unknown discrete-time model for each pinned node $i = 1, 2, \ldots, l$, defined as

$$\mathbf{x}_{i,k+1} = \hat{f}_{ik}(\mathbf{x}_{i,k}) + \hat{g}_{ik}(\mathbf{x}_{i,k})\mathbf{d}(k) + \mathbf{u}_{i,k}, \qquad (6.31)$$

where the vector functions $\hat{f}_{ik}(\cdot)$ and $\hat{g}_{ik}(\cdot)$, and disturbance $\mathbf{d}(k)$ are unknown. Assume that all states $\mathbf{x}_i(k)$ are available at the sampling instants. Let system (6.31) be approximately modelled by an RHONN, described by

$$\mathbf{x}_{i,k+1} = W_i^{*T}(k)\boldsymbol{\phi}_i(\mathbf{x}_{i,k}) + \mathbf{u}_{i,k} + \epsilon,$$

where $W_i^*(k) \in \mathbb{R}^{L \times n}$ are the ideal optimal constant weight matrices; $\boldsymbol{\phi}_i(\cdot) \in \mathbb{R}^L$ are defined as

$$\boldsymbol{\phi}_i(\mathbf{x}_{i,k}) = \begin{bmatrix} \prod_{j \in I_1} \zeta_{ij}^{d_{ij}(1)} \\ \vdots \\ \prod_{j \in I_L} \zeta_{ij}^{d_{ij}(L)} \end{bmatrix},$$

where d_{ij} are non-negative integers; L is the respective number of high-order connections; $\{I_1, I_2, \ldots, I_L\}$ is a collection of non-ordered subset of $\{1, 2, \ldots, n\}$; and ζ_i is given by

$$
\zeta_i = \begin{bmatrix} \zeta_{i_1} \\ \vdots \\ \zeta_{i_n} \end{bmatrix} = \begin{bmatrix} S(x_1) \\ \vdots \\ S(x_n) \end{bmatrix},
$$

where $S(\cdot)$ is a hyperbolic tangent function. Finally, let ϵ be the modeling error, so that

$$
\|\epsilon\| \leq \epsilon_N,
$$

where ϵ_N is a known bound.

A discrete-time RHONN identifier with a series-parallel structure is implemented to identify (6.31), as

$$
\chi_{i,k+1} = W_{i,k}^T \phi_i(\mathbf{x}_{i,k}) + \mathbf{u}_{i,k}, \tag{6.32}
$$

where $\chi_i = [\chi_{i_1}, \chi_{i_2}, \ldots, \chi_{i_n}]^T \in \mathbb{R}^n$ is the neuron state vector, and $W_i \in \mathbb{R}^{L \times n}$ is the adjustable weights vector.

The Extended Kalman Filter (EKF) learning algorithm is used to on-line adjust the neural weights, which determines the values W_i to minimize the identification error.

6.3.2 Discrete-time inverse optimal controller

Consider an affine-in-the-input discrete-time nonlinear form of the identified pinned nodes, formulated as

$$
\begin{aligned}
\mathbf{x}_{i,k+1} &= W_{i,k}^T \phi_i(\mathbf{x}_{i,k}) + \mathbf{u}_{i,k}, \\
&\triangleq f_{i_k}(\mathbf{x}_{i,k}) + g_{i_k}(\mathbf{x}_{i,k})\mathbf{u}_{i,k}.
\end{aligned}
$$

The discrete-time inverse optimal control law as in [15] is employed to guarantee trajectory tracking for the pinned nodes:

$$\mathbf{u}_{i,k} = \alpha(\mathbf{x}_{i,k}) = -\left(R_k + P_1(\mathbf{x}_{i,k})\right)^{-1} P_2(\mathbf{x}_{i,k}), \qquad (6.33)$$

where

$$P_1(\mathbf{x}_{i,k}) = \frac{1}{2} g_{i_k}^T(\mathbf{x}_{i,k}) \, P \, g_{i_k}(\mathbf{x}_{i,k}),$$

and

$$P_2(\mathbf{x}_{i,k}) = g_{i_k}^T(\mathbf{x}_{i,k}) \, P \left(f_{i_k}(\mathbf{x}_{i,k}) - \mathbf{x}_{r,k+1}\right).$$

The complete stability analysis of this controller is explained in [15].

6.3.3 Simulation results

To illustrate the effectiveness of the controller, consider a complex network consisting of 10 nodes. This network includes different chaotic dynamics as follows: the Chen system (Nodes 1 and 2) [3], Lorenz system (Nodes 3 and 4) [9], Lü system (Nodes 5 and 6) [10], Rössler system (Nodes 7 and 8) [16], and Chua's system (Nodes 9 and 10) [11]. The adjacency matrix A is given as

$$A = \begin{bmatrix}
-7 & 1 & 1 & 1 & 0 & 1 & 0 & 1 & 1 & 1 \\
1 & -5 & 1 & 1 & 0 & 0 & 0 & 1 & 1 & 0 \\
1 & 1 & -5 & 1 & 1 & 0 & 1 & 0 & 0 & 0 \\
1 & 1 & 1 & -4 & 0 & 0 & 1 & 0 & 0 & 0 \\
0 & 0 & 1 & 0 & -2 & 1 & 0 & 0 & 0 & 0 \\
1 & 0 & 0 & 0 & 1 & -2 & 0 & 0 & 0 & 0 \\
0 & 0 & 1 & 1 & 0 & 0 & -2 & 0 & 0 & 0 \\
1 & 1 & 0 & 0 & 0 & 0 & 0 & -2 & 0 & 0 \\
1 & 1 & 0 & 0 & 0 & 0 & 0 & 0 & -2 & 0 \\
1 & 0 & 0 & 0 & 0 & 0 & 0 & 0 & 0 & -1
\end{bmatrix}.$$

The coupling strengths are $c_{ij} > 100$, and $\mathbf{\Gamma} = diag(1,1,1)$. The control objective is to track a desired trajectory described by the Arneodo system [2]

Only one controller is applied, at Node 1, and the objective is to track the desired trajectory. This controller is defined as (6.33). To model the pinned node, the following RHONN identifier, formulated based on the measurements of $\mathbf{x}_1(k) = [x_{1_1}(k), x_{1_2}(k), x_{1_3}(k)]^T$, is used:

$$
\begin{aligned}
x_{1_1,k+1} &= W_1^T \phi_1(\mathbf{x}_{1,k}) + u_{1,k}, \\
x_{1_2,k+1} &= W_2^T \phi_2(\mathbf{x}_{1,k}) + u_{2,k}, \\
x_{1_3,k+1} &= W_3^T \phi_3(\mathbf{x}_{1,k}) + u_{3,k},
\end{aligned}
$$

with

$$
\begin{aligned}
\phi_1(\mathbf{x}_{1,k}) &= [S(x_{1_1,k}), S(x_{1_2,k}), S^2(x_{1_1,k}), \\
& \quad S^3(x_{1_1,k}), S^4(x_{1_1,k})]^T, \\
\phi_2(\mathbf{x}_{1,k}) &= [S(x_{1_2,k}), S(x_{1_1,k}), S(x_{1_1,k})S(x_{1_3,k}), \\
& \quad S(x_{1_2,k})S(x_{1_3,k}), S^2(x_{1_2,k})]^T, \\
\phi_3(\mathbf{x}_{1,k}) &= [S^3(x_{1_3,k}), S(x_{1_1,k})S(x_{1_2,k}), \\
& \quad S(x_{1_1,k})S(x_{1_2,k}), S^2(x_{1_3,k})]^T,
\end{aligned}
$$

where $S(\cdot) = tanh(\cdot)$.

Simulations are performed as follows. From $t = 0s$ to $t = 5s$, the network runs without interconnections and without control actions ($c_{ij} = 0, \mathbf{u}_1 = 0$). Starting from $t = 5s$, the coupling strengths are selected as $c_{ij} > 100$, the control law is applied, and the complex network is forced to track the desired trajectory. Figs. 6.7(a) and 6.7(b) exhibit the evolutions of the network states and their average, respectively. Fig. 6.8(a) presents the control input signal $\mathbf{u}_i(t)$ applied to the pinned node. The tracking error $\mathbf{e}_i$ of the pinned node is presented in Fig. 6.8(b). Figs. 6.8(c) and 6.8(d) exhibit the identification error

ε_i and the weights evolution for Node 1, respectively. These results illustrate that the control scheme achieve the trajectory tracking objective.

6.4 Conclusions

- A new control strategy is introduced for trajectory tracking on uncertain complex networks, based on pinning control and using a recurrent high-order neural network for identification. This control scheme utilizes the inverse optimal control technique, which guarantees the whole network to track the desired trajectories, even in the presence of randomly time-varying coupling strengths. The control scheme is evaluated via simulation examples, demonstrating its good performance and effectiveness.

- The trajectory tracking issue for complex networks with nonidentical nodes by using sampled-data pinning controller is analyzed. A new criterion is presented, combining the Lyapunov–Krasovskii approach and the V-stability tool. The controller is constituted by an RHONN identifier trained with an EKF to perform pinned-node identification, and the inverse optimal control technique is employed to achieve the objective that the network state tracks the desired reference signal. Simulation results illustrate that for a network even consisting of nodes with different chaotic dynamics, the sampled-data control scheme works very well.

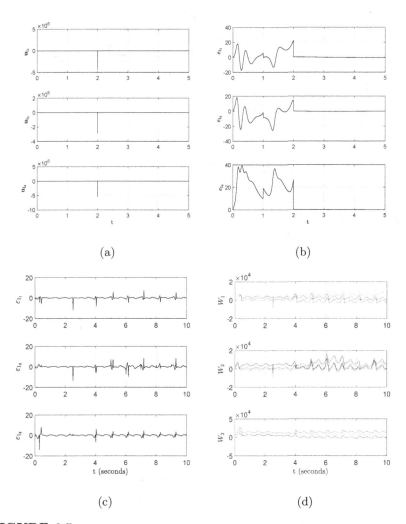

FIGURE 6.5

Results for performance of the NN-based inverse optimal pinning control law:
(a) Signal of the control law $u_{1_i}(t)$, $i = 1, 2, 3$; (b) Tracking error $\mathbf{e}_1$; (c)
Identification error $\boldsymbol{\varepsilon}_1$; (d) Weights evolution at Node 1 with identification.

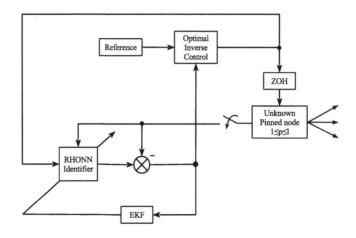

FIGURE 6.6
The sampled-data pinning control scheme.

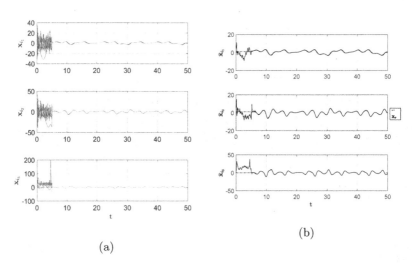

(a)

(b)

FIGURE 6.7
Results of scale-free network model with five types of nodes, which are
described by the Chen system (Nodes 1 and 2), Lorenz system (Nodes 3 and
4), Lü system (Nodes 5 and 6), Rössler system (Nodes 7 and 8) , and Chua's
system (Nodes 9 and 10): (a) The evolution of network states; (b) Average
trajectory of network states, $\bar{\mathbf{x}}_i$ (- continuous line), reference trajectory, $\mathbf{x}_r$
($-\cdot$ dashed line)

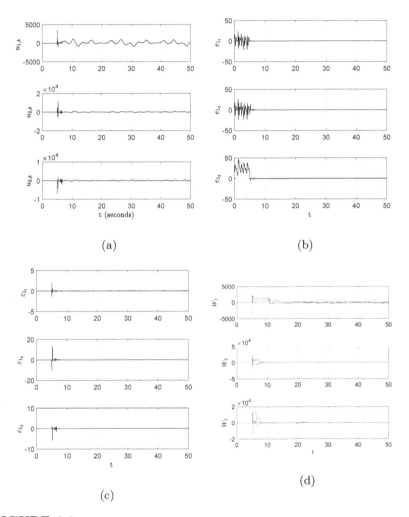

FIGURE 6.8
Results for performance of the discrete-time NN-based inverse optimal pinning control law: (a) Signal of the control law $u_{1_i}(t)$, $i = 1, 2, 3$; (b) Tracking error $\mathbf{e}_1$; (c) Identification error $\boldsymbol{\varepsilon}_1$; (d) Weights evolution at Node 1 with identification.

Bibliography

[1] A. Y. Alanis, E. N. Sanchez, A. G. Loukianov, and E. A. Hernandez. Discrete-time recurrent high order neural networks for nonlinear identification. *Journal of the Franklin Institute*, 347(7):1253–1265, 2010.

[2] A. Arneodo, P. Coullet, E. Spiegel, and C. Tresser. Asymptotic chaos. *Physica D: Nonlinear Phenomena*, 14(3):327–347, 1985.

[3] G. Chen and T. Ueta. Yet another chaotic attractor. *International Journal of Bifurcation and Chaos*, 9(7):1465–1466, 1999.

[4] H. K. Khalil. *Nonlinear Systems,* 2nd ed. Upper Saddle River, NJ, USA: Prentice Hall, 1996.

[5] E. B. Kosmatopoulos, M. M. Polycarpou, M. A. Christodoulou, and P. A. Ioannou. High-order neural network structures for identification of dynamical systems. *IEEE Transactions on Neural Networks*, 6(2):422–431, 1995.

[6] M. Krstic and Z. H. Li. Inverse optimal design of input-to-state stabilizing nonlinear controllers. *IEEE Transactions on Automatic Control*, 43(3):336–350, 1998.

[7] X. Li and G. Chen. Synchronization and desynchronization of complex dynamical networks: An engineering viewpoint. *IEEE Transactions on Circuits and Systems I: Fundamental Theory and Applications*, 50(11):1381–1390, 2003.

[8] X. Li, X. Wang, and G. Chen. Pinning a complex dynamical network to its equilibrium. *IEEE Transactions on Circuits and Systems I: Regular Papers*, 51(10): 2074–2087, 2004.

[9] E. N. Lorenz. Deterministic nonperiodic flow. *Journal of the Atmospheric Sciences*, 20(2):130–141, 1963.

[10] J. Lü and G. Chen. A new chaotic attractor coined. *International Journal of Bifurcation and Chaos*, 12(3):659–661, 2002.

[11] T. Matsumoto, L. Chua, and M. Komuro. The double scroll. *IEEE Transactions on Circuits and Systems*, 32(8):797–818, 1985.

[12] G. A. Rovithakis and M. A. Christodoulou. *Adaptive Control with Recurrent High-Order Neural Networks*. London, UK: Springer-Verlag, 2000.

[13] E. N. Sanchez, A. Y. Alanís, and A. G. Loukianov. *Discrete-Time High Order Neural Control: Trained with Kalman Filtering*. Berlin, Germany: Springer, 2008.

[14] E. N. Sanchez and D. I. Rodriguez. Inverse optimal pinning control for complex networks of chaotic systems. *International Journal of Bifurcation and Chaos*, 25(2):1550031, 2015.

[15] E. N. Sanchez and F. Ornelas-Tellez. *Discrete-Time Inverse Optimal Control for Nonlinear Systems*. Boca Raton, FL, USA: CRC Press, 2017.

[16] O. E. Rössler. An equation for continuous chaos. *Physics Letters A*, 57(5):397–398, 1976.

Part IV

Applications

7

Pinning Control for the p53-Mdm2 Network

Gene p53 regulates the cellular response to genotoxic damage and prevents carcinogenic events. Theoretical and experimental studies state that the p53-Mdm2 network constitutes the core module of regulatory interactions activated by cellular stress induced by a variety of signaling pathways. In this chapter, a strategy to control the p53-Mdm2 network regulated by p14ARF is developed, based on the pinning control technique. Pinned nodes are selected on the basis of their importance level in a topological hierarchy, their degree of connectivity within the network, and the biological role they perform. Two cases are considered. For the first case, the oscillatory pattern under gamma-radiation is recovered; for the second case, increased expression of p53 level is taken into account. For both cases, the control law is applied to p14ARF (pinned node based on a virtual leader methodology), and overexpressed Mdm2-mediated p53 degradation is considered as carcinogenic initial behavior.

7.1   p53-Mdm2 Model Regulated by p14ARF

Gene regulatory networks present responses to DNA damage such as cell cycle arrest, DNA repair, senescence, and apoptosis. Among the main regulators of these responses, tumor suppressor protein p53 ($TP53$) has been recognized as the "guardian of the genome" and is a key component of cellular responses to genotoxic stress [22]. Gene p53 regulates the cellular response to genotoxic damage and prevents tumorigenesis by post-translational modifications and gene transactivation [1]. Without cellular stress, p53 remains inactive and latent due to targeted degradation by the protein E3 ubiquitin-protein ligase Mdm2 (from $MDM2$ proto-oncogene). Mdm2 binds to p53 and marks it for proteasome degradation, preventing p53 accumulation in the nucleus and its transcriptional activity [26]. In this way, the activity of p53 and its negative regulation by Mdm2 are widely recognized as one of the main regulatory mechanisms in genotoxic stress response [37]. The p53-Mdm2 network models have been studied in [43, 14, 41, 45, 46, 16], which describe different patterns of response as promotion of transcriptional activities, post-translational modifications, component interactions and degradation rates.

Another important regulator of the p53-Mdm2 network is the tumor suppressor p14ARF (Alternate Reading Frame), one product of the $CDKN2A$ gene. It is known that p14ARF inhibits Mdm2-dependent p53 degradation, through Mdm2-p14ARF complex formation [48]. Thus, in response to genotoxic stress induced by gamma-radiation, p14ARF binds directly to Mdm2, leading to an inhibition of Mdm2-mediated p53 ubiquitination and degradation, which increases p53 levels. These events are coupled with downstream signaling pathways, promoting behaviors such as cell cycle arrest, DNA repair, senescence, and apoptosis induction [48, 37, 30].

The p53-Mdm2 network is key for determining cell behavior in response to cellular stress, such as DNA damage induced by gamma-radiation [39, 17], which can start a program of p53-dependent consequences such as cell cycle arrest, DNA repair, senescence or apoptosis induction [43, 14, 41, 45, 46, 16]. In Fig. 7.1, the interaction network between p53, Mdm2, and p14ARF is shown for an individual cell model in two cell compartments: nucleus and cytoplasm. This model includes DNA damage induced by gamma-radiation, which generates the p53 activation and the transactivation of both $TP53$ and $MDM2$, among other transactivated genes, which are not considered in this model. Initially, the translocation process of mRNAs to the cytoplasm and its subsequent translation into proteins takes place; then, proteins are transported back to the nucleus. While p53 remains at high levels, $Mdm2_{nuclear}$ reduces its concentration levels, and vice versa, thus producing an oscillatory pattern. $Mdm2_{cytoplasmic}$ moves to the nucleus at a constant rate, ignoring all other possible behaviors. The production and degradation rates of p14ARF remain constant [23].

The following features of the p53-Mdm2 network regulated by p14ARF are presented:

1. The p53-Mdm2 network is one of the mostly explored biological mechanisms that provides adequate information.

2. The biological context requires different scenarios depending on p53 and failure in the network of p53-Mdm2-p14ARF, which can be represented by equations for outputting a desired reference that have a biological explanation, such as regulation mechanisms in gene expression.

3. The biological responses of p53 in exploring are related to the ability of generating cell cycle arrest, DNA repair, senescence, and/or apoptosis, especially the last one since it has a key role in the tumor suppressor response.

4. The model assumes the response to an ionizing radiation stimulus with p53-dependent responses, which presents an oscillatory pattern

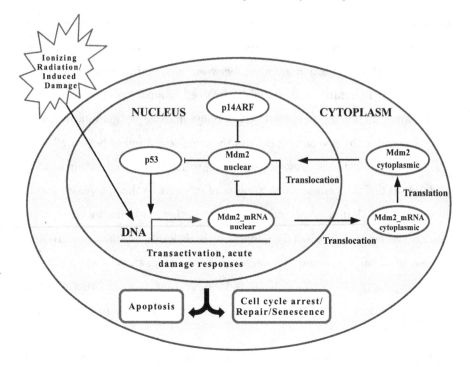

FIGURE 7.1
Schematic model of p53 network including $Mdm2_{nuclear}$ inhibition by
p14ARF. p53 stabilization leads to transcriptional activity, inducing
$Mdm2_{mRNA}$ expression. $Mdm2_{cytoplasmic}$ is transported to the nucleus.
p14ARF is also upregulated under genotoxic stress, where binds to
$Mdm2_{nuclear}$ and causing its suppression. p53 downstreaming possible
responses include cell cycle arrest, DNA repair, senescence, and/or apoptosis.

due to variation between p53 and its inhibitor Mdm2. The main

objective is to simulate the p53 suppressed response when Mdm2

is overexpressed, which has been documented in several types of

tumors.

7.2 Mathematical Description

Taken from [23], based on the principle of mass-action and the saturable transcription kinetics, the p53-Mdm2 network regulated by p14ARF without control action is mathematically described as follows:

$$\frac{d[p53]}{dt} = k_p - k_1[p53][Mdm2_{nuclear}] - d_p[p53], \quad (7.1)$$

$$\frac{d[Mdm2\_mRNA_{nuclear}]}{dt} = k_m + k_2\frac{[p53]^{1.8}}{K_D^{1.8} + [p53]^{1.8}}$$

$$- k_0[Mdm2\_mRNA_{nuclear}], \quad (7.2)$$

$$\frac{d[Mdm2\_mRNA_{cytoplasmic}]}{dt} = k_0[Mdm2\_mRNA_{nuclear}]$$

$$- d_{rc}[Mdm2\_mRNA_{cytoplasmic}], \quad (7.3)$$

$$\frac{d[Mdm2_{cytoplasmic}]}{dt} = k_T[Mdm2\_mRNA_{cytoplasmic}]$$

$$- k_i[Mdm2_{cytoplasmic}], \quad (7.4)$$

$$\frac{d[Mdm2_{nuclear}]}{dt} = k_i[Mdm2_{cytoplasmic}] - d_{mn}[Mdm2_{nuclear}]^2$$

$$- k_3[Mdm2_{nuclear}][p14ARF], \quad (7.5)$$

$$\frac{d[p14ARF]}{dt} = k_a - d_a[p14ARF]$$

$$- k_3[Mdm2_{nuclear}][p14ARF]. \quad (7.6)$$

The deterministic model ((7.1)–(7.6)) includes: p53 production and degradation (equation (7.1)), $Mdm2\_mRNA_{nuclear}$ basal transcription (p53-dependent and Mdm2-independent production) in equation (7.2); in (equation (7.3)) $Mdm2\_mRNA_{cytoplasmic}$ transport from nucleus to cytoplasm, $Mdm2\_mRNA_{cytoplasmic}$ translation rate and $Mdm2_{cytoplasmic}$ protein transport to nucleus (equation (7.4)), $Mdm2_{cytoplasmic}$ decay through $Mdm2_{nuclear}$-p14ARF complex, which removes Mdm2 and stops $Mdm2_{nuclear}$ ubiquitination rate (equation (7.5)), p14ARF production and p14ARF decay

Parameter	Description	Value
k_p	p53 production	0.5 proteins/s
k_1	Mdm2-dependent p53 degradation	9.963×10^{-6}/s
d_p	p53 decay	1.925×10^{-5}/s
k_m	p53-independent Mdm2 production	1.5×10^{-3} RNA/s
k_2	p53-dependent Mdm2 production	1.5×10^{-2}/s
K_D	Dissociation constant in the promoter region	740 proteins
k_0	RNA transport from nucleus to cytoplasm	8.0×10^{-4}/s
d_{rc}	Mdm2_mRNA decay in cytoplasm	1.444×10^{-4}/s
k_T	Translation rate	1.66×10^{-2} proteins/s
k_i	Protein transport from cytoplasm to nucleus	9.0×10^{-4}/s
d_{mn}	Mdm2 autoubiquitination	1.66×10^{-7}/s
k_3	Mdm2$_{nuclear}$-p14ARF complex formation rate	9.963×10^{-6}/s
K_a	p14ARF production	0.5 proteins/s
d_a	p14ARF decay	3.209×10^{-5}/s

TABLE 7.1
Parameters values for p53-Mdm2 network regulated by p14ARF.

in nucleus compartment (equation (7.6)). The parameter values used are presented in Table 7.1.

7.3 Pinning Control Methodology

7.3.1 Problem formulation

Consider a general network consisting of N nodes with nonlinear couplings, where each node is a scalar nonlinear dynamical system, which represents genes, concentrations of RNAs and proteins, given by

$$\dot{x}_i = f_i(x_i) + h_i(t, x_1, x_2, \ldots, x_N), \quad i = 1, 2, \ldots, N, \qquad (7.7)$$

where $x_i \in \mathbb{R}$ is the state of node i, for $i = 1, 2, \ldots, N$; $f_i : \mathbb{R} \mapsto \mathbb{R}$ represents the self-dynamics of node i related to the degradation process of RNA and proteins; $h_i : \mathbb{R}^N \mapsto \mathbb{R}$ is the nonlinear coupling function between nodes, associated with the changes of x_i due to transcription, translation, repression, activation or other interaction processes, and N is the number of network nodes.

The control objective is that (7.7) tracks a desired output trajectory, given by

$$y = y_r(t).$$

To achieve this objective, local feedback controllers are applied to a reduced number of network nodes, according to the pinning control methodology [44, 25, 9, 40]. This methodology is composed of two parts: the first one, pinned nodes l are selected to apply control actions as in (7.8) , where $1 \leq l \leq N$ and l can be as small as one. The second one is the remained network nodes $(N - l)$ without control action as in (7.9). Thus, the controlled network can be written as

$$\dot{x}_i = f_i(x_i) + h_i(t, x_1, x_2, \ldots, x_N) + g_i(x_i)u_i, \quad i = 1, 2, \ldots, l. \qquad (7.8)$$

$$\dot{x}_i = f_i(x_i) + h_i(t, x_1, x_2, \ldots, x_N), \quad i = l+1, l+2, \ldots, N. \qquad (7.9)$$

where $g_i : \mathbb{R} \mapsto \mathbb{R}$ is a nonlinear function of the node state i, for $i = 1, 2, \ldots, l$, and u_i denotes the control on the node $i \in l$.

In the present section, u_i in (7.8) is a local positive discontinuous feedback control law, described by

$$u_i = \begin{cases} 1 + K_i(1 - e_i), & \text{if} \quad |\varphi_i| < 1, \\ 1 + K_i(1 - sign(e_i)), & \text{if} \quad |\varphi_i| > 1, \end{cases} \qquad (7.10)$$

where K_i is a positive control gain selected by the designer, e_i is the tracking error between the desired output trajectory $(y_r(t))$ and the controlled state (x_i), given by

$$e_i = (x_i - y_r(t)), \qquad (7.11)$$

with $\varphi_i = \frac{e_i}{S_i}$ being an auxiliary variable to reject chattering effect caused by $sign(\cdot)$ (signum function extracts the sign of a real number) [38], and S_i a signal filter given by

$$\dot{S}_i = -\alpha_i S_i + \omega_i, \quad i = 1, 2, \ldots, l, \qquad (7.12)$$

where α_i and ω_i are positive gains to be selected.

7.3.2   p14ARF Pinned node

To select the pinned nodes, the virtual leader methodology presented in [34, 24] is used. The methodology consists in analyzing the interactions between proteins presented in Fig. 7.1 using the mathematical model ((7.1)–(7.6)). The nodes that affect directly or indirectly the dynamical behavior of everyone else, are candidates as pinned nodes. In this sense, the spanning tree of the p53-Mdm2 network regulated by p14ARF is developed, as shown in Fig. 7.2;

based on this analysis, p14ARF is the adequate biological selection as the pinned node.

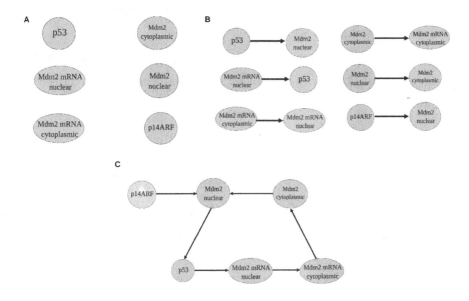

FIGURE 7.2

Spanning tree process of the p53-Mdm2 network regulated by p14ARF. In (**A**), proteins and mRNA of the network are presented. In (**B**), based on the mathematical model ((7.1)–(7.6)) with a biological analysis, proteins and mRNA are illustrated with their dependent molecule. In (**C**), the spanning tree, where P14ARF do not have direct dependence (virtual leader) from other protein/mRNA, is presented.

From (7.6), the differential equation for p14ARF (pinned node) is defined by three cellular processes, as follows:

$$\frac{d[p14ARF]}{dt} = \overbrace{k_a}^{\text{Production}} \overbrace{-d_a[p14ARF]}^{\text{Degradation}} \overbrace{-k_3[Mdm2_{nuclear}][p14ARF]}^{Mdm2_{nuclear}-p14ARF \text{ complex formation}} .$$

In order to control the p53-Mdm2 network, it is necessary to increase the p14ARF concentration levels to regulate $Mdm2_{nuclear}$ production. As can be seen, degradation process and $Mdm2_{nuclear} - p14ARF$ complex formation have a negative sign; while the production process has a positive sign. Due to this fact, it is possible to modify the p14ARF production (K_a) process by adding the control law (7.10).

Thus, considering equations (7.8) and (7.9), the p53-Mdm2 network regulated by p14ARF with control action is mathematically described as follows:

$$\frac{d[p14ARF]}{dt} = k_a u_1 - d_a[p14ARF]$$
$$- k_3[Mdm2_{nuclear}][p14ARF], \qquad (7.13)$$

$$\frac{d[p53]}{dt} = k_p - k_1[p53][Mdm2_{nuclear}] - d_p[p53], \quad (7.14)$$

$$\frac{d[Mdm2\_mRNA_{nuclear}]}{dt} = k_m + k_2 \frac{[p53]^{1.8}}{K_D^{1.8} + [p53]^{1.8}}$$
$$- k_0[Mdm2\_mRNA_{nuclear}], \qquad (7.15)$$

$$\frac{d[Mdm2\_mRNA_{cytoplasmic}]}{dt} = k_0[Mdm2\_mRNA_{nuclear}]$$
$$- d_{rc}[Mdm2\_mRNA_{cytoplasmic}], \qquad (7.16)$$

$$\frac{d[Mdm2_{cytoplasmic}]}{dt} = k_T[Mdm2\_mRNA_{cytoplasmic}]$$
$$- k_i[Mdm2_{cytoplasmic}], \qquad (7.17)$$

$$\frac{d[Mdm2_{nuclear}]}{dt} = k_i[Mdm2_{cytoplasmic}] - d_{mn}[Mdm2_{nuclear}]^2$$
$$- k_3[Mdm2_{nuclear}][p14ARF]. \qquad (7.18)$$

The p53-Mdm2 network regulated by p14ARF without control action ((7.1)–(7.6)) or with control action ((7.13)–(7.18)) are simulated using Matlab/Simulink and the fourth-order Runge-Kutta integration method with a fixed step size of 1×10^{-3} as further discussed respectively in the following sections.

7.4 Behaviors of the p53-Mdm2 Network Regulated by p14ARF without Control Action

7.4.1 p53-Mdm2$_{nuclear}$ oscillatory pattern

As displayed in Fig. 7.3, the p53-Mdm2 network presents an oscillatory pattern under gamma-radiation, for a lapse of 48 hours, using the parameter values shown in Table 7.1. This response is due to post-translational modifications of p53 and the negative interactive loop of Mdm2-mediated ubiquitination, according to [10, 14]; this pattern has not been observed for all cell types and requires wild-type genes [21, 23].

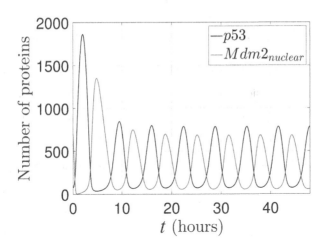

FIGURE 7.3

p53-Mdm2$_{nuclear}$ oscillatory pattern under gamma-radiation within a lapse of 6.64 hours.

Levels of p53 and Mdm2$_{nuclear}$ proteins present oscillatory behavior, caused by pulses resulting from p53 activation, p53-dependent transactivation, Mdm2 production, Mdm2$_{nuclear}$ sequestration by p14ARF, and Mdm2-mediated ubiquitination. This process allows the cell to respond to double-strand breaks (DSBs) in DNA induced by gamma radiation, which correlates with the number of p53 pulses in individual cells [21, 14]. For the model used in this application, there are regulatory factors, which have not been considered, as the interaction of other potentially relevant genes transactivated by p53 (nearly 100 genes in pathways such as cell cycle arrest, DNA repair, senescence, and apoptosis [35]). One of the simpler explanations for the p53 oscillatory pattern is due to repeated activation of ATM (Ataxia Telangiectasia Mutant), which dissociates the p53-Mdm2 complex and stabilizes an increase in p53 levels. These changes are driven by persistent DNA damage induced by radiation [21, 14]. Therefore, recovering the normal pattern response to DNA damage by the p53-Mdm2 network is fundamental for tumor suppression.

7.4.2   Mdm2$_{nuclear}$ overexpression and p53 downregulation

Fig. 7.4 illustrates that p53-Mdm2-dependent affinity is altered, producing a Mdm2$_{nuclear}$ overexpression when the parameters k_1 and k_2 are set to values five and ten times larger than the original ones, respectively. This behavior is reported in a variety of human soft tissue tumors and in hematological malignancies, as discussed in [28, 27, 11, 5, 33]. In human tumors, Mdm2 overexpression can inhibit p53 normal regulatory activities and induce a loss of growth-inhibitory signals in cytostatic and apoptotic responses, which may favor a carcinogenic process.

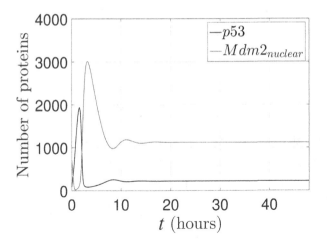

FIGURE 7.4

Mdm2$_{nuclear}$ overexpression and p53 downregulation with $k_1 = 1.9926 \times 10^{-6}$/s and $k_2 = 45 \times 10^{-3}$/s; once Mdm2$_{nuclear}$ is overexpressed, p53 is inhibited. It can be conjectured that Mdm2$_{nuclear}$ deregulation will lead to oncogenic behavior through p53 suppression.

There are gene abnormalities in tumors, carcinogenesis driven by viral infections, and other mechanisms that contribute to inactivate p53 functions and its signaling outcomes [36, 6, 33]. For example, Mdm2 overexpression leads to p53 downregulation, contributing to losses of tumor suppressor activities. Mdm2 overexpressed is a hallmark in several types of cancer [27, 11, 33]. As reviewed in [33], Mdm2 is overexpressed in liposarcomas, osteosarcomas, testicular germ cell tumors, embryonic carcinomas, brain tumors (including glioblastomas and astrocytomas), hematological malignancies, bladder cancer, breast cancer, colorectal cancer, among others. In this sense, the number of Mdm2 abnormalities is highly variable. Moreover, not all samples from the same type of tumor showing Mdm2 overexpression [5] report other mechanisms that promote Mdm2 overexpression. Whereas a single nucleotide polymorphism (SNP309) in the MDM2 promoter increases the affinity of the transcriptional activator Sp1, resulting in higher levels of Mdm2 mRNA and Mdm2 translation rate. This behavior illustrates p53 downregulation, resulting in a decreased response to DNA damaging agents and acceleration of tumorigenesis.

7.4.3 Increased expression of p53 levels

Fig. 7.5 displays an increased expression of p53 levels when the parameters K_a and d_{rc} are set to values ten and one hundred times larger than the original ones, respectively. This p53 dynamical behavior is related to downstream genes involved in signaling pathway, which can produce cell cycle arrest, DNA repair, senescence and/or apoptotic response [13, 19, 20, 2, 32].

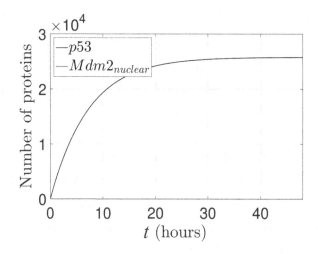

FIGURE 7.5

Increased expression of p53 levels response under gamma-radiation induction with $k_a = 5$ proteins/s and $d_{rc} = 1.444 \times 10^{-2}$/s. It can be conjectured that p53 upregulation will lead to cell cycle arrest, DNA repair, senescence and/or apoptotic response.

The stabilization and accumulation of p53 levels are part of the DNA damage response to maintain genome integrity. One of the possible response outputs of a fully functional p53 pathway is the induction of apoptosis generated by DNA damage. For apoptosis induction by radiation in certain tissues, a cascade of signaling is generated through pro-apoptotic proteins, such as response mediated by ATM protein [3]. ATM leads to p53 stabilization, modifying the interaction capabilities with Mdm2 [13]. With the activation of p53, its degradation is limited, and p53 levels increase [29]. Downstream to ATM/p53 activation, an apoptosis program is carried out by a set of proteins, such as Bim, Puma, Bid, Bmf, Bad, Bik, Noxa, and Hrk, whereby Puma and Noxa can be directly regulated with p53 overexpression by gamma-radiation [42, 8]. This set of proteins can bind and block survival proteins such as Bcl-2, which release death effectors like Bax and Bak. These effectors can lead to a change in the permeability of the outer mitochondria membrane. Furthermore, they can participate in the cell dismantling coupled with caspases [15]. There are other possible outcomes for the p53 activation pathway and p53 independent responses to DNA damage by gamma-radiation, which can lead the cell to cell cycle arrest, initiate DNA repair, or perform senescence. Experimental evidence indicates that the cellular level of p53 can dictate the response of the cell, such that lower levels of p53 result in arrest, whereas higher level results in apoptosis [7, 32].

7.4.4 Sensitivity analysis for p14ARF production

This analysis is done on the basis of p14ARF production (K_a) value variation effects on p53 and Mdm2$_{nuclear}$, respectively as can be seen in Fig. 7.6. Sensitivity analysis [12, 18] determines the K_a values for which the network cannot achieve desired behaviors (0–1.5 proteins/s); the K_a value to reach p53-Mdm2$_{nuclear}$ oscillatory pattern (1.6–9.5 proteins/s), and the K_a value to generate an increased expression of p53 with Mdm2$_{nuclear}$ downregulation (9.6–50 proteins/s) are determined from Fig. 7.6.

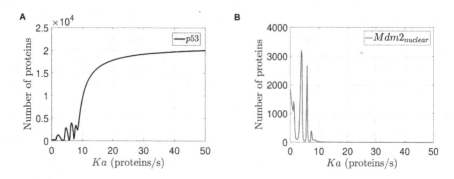

FIGURE 7.6
Sensitivity analysis for K_a. In (**A**), p53 response to K_a variations is illustrated. In (**B**), $Mdm2_{nuclear}$ response to K_a variations is presented. For both cases K_a varies between 0 proteins/s to 50 proteins/s for all network proteins, p53 and $Mdm2_{nuclear}$ are selected because they are the desired output in the network.

7.5   Behaviors of the p53-Mdm2 Network Regulated by p14ARF with Control Action

To illustrate the p53 and $Mdm2_{nuclear}$ behavior under pinning control actions, two cases are considered: (1) to restore an oscillatory pattern under gamma-radiation, and (2) to achieve an increased expression of p53 level. For a 24 hours lapse, the network runs without any control action and presents overexpressed Mdm2-mediated p53 degradation as carcinogenic initial behavior for both cases ([28, 27, 11, 5, 33]), which is displayed in Fig. 7.4.

7.5.1   Case 1: Restoration of an oscillatory pattern under gamma-radiation

For the first case, by considering p14ARF production ($k_a u_1$) in a range between 0 and 1.5 proteins/s, the proposed controller is turned on after the 24 hours initial lapse; however, the network can not achieve the oscillatory pattern as can be seen in Fig. 7.7(A). Due to the low value of $k_a u_1$, the pinning control technique cannot achieve the desired behavior. Otherwise, with $(k_a u_1)$ in a range between 1.6–9.5 proteins/s, the controller forces the network to gradually track the oscillatory pattern as can be seen in Fig. 7.7(B).

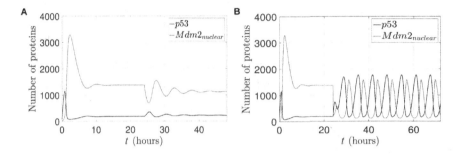

FIGURE 7.7
The scenario simulations to achieve restoration of an oscillatory pattern under gamma-radiation in the p53-Mdm2 network regulated by p14ARF. For both tests it is possible to observe overexpressed Mdm2-mediated p53 degradation as carcinogenic initial behavior at first 24 hours. In (**A**), control action with $(k_a u_1)$ in a range between 0–1.5 proteins/s, the oscillatory pattern under gamma-radiation is not achieved (24–48 hours), and likewise in (**B**) oscillatory pattern under gamma-radiation by pinning control (24–72 hours) with $(k_a u_1)$ in a range between 1.6–9.5 proteins/s is achieved.

The oscillatory pattern behavior [14, 45, 4] induced by means of the pinning control technique is illustrated in Fig. 7.7, with the purpose of restoring normal network behavior in the presence of oncogenic overexpressed Mdm2; this overexpression avoids normal regulatory activities due to a suppressed wild-type p53. The pinning control technique (7.10) is located at the production of p14ARF node with $K_i = 100$, which allows to regenerate an oscillatory pattern and guarantees that p14ARF production $k_a u_1$ achieves a range between 1.6 and 9.5 proteins/s. Enhanced p14ARF production induces a decrease in $Mdm2_{nuclear}$, which in turn increased p53 levels. This approach can help analyze multiple interaction mechanisms, and to induce different cell reprogramming responses. Therefore, this approach motivates future research on the interdependencies of cellular networks and new ways for treatment designs in tumor suppressor networks.

7.5.2 Case 2: Achievement of a p53 level increased expression

For the second case, by considering p14ARF production $(k_a u_1)$ in a range between 0 and 1.5 proteins/s, the controller is turned on; however, the network cannot achieve increased expression of p53 as can be seen in Fig. 7.8(A). Due to the low value of $k_a u_1$, the pinning control technique cannot achieve the desired behavior. Otherwise, with $(k_a u_1)$ in a range between 9.6 and 50 proteins/s, the controller again is turned on, and the system gradually tracks the increased expression of p53 levels as can be seen in Fig. 7.8(B).

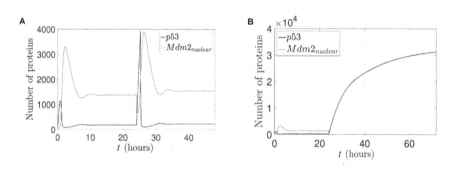

FIGURE 7.8

The scenario simulations to achieve p53 level increased expression in the p53-Mdm2 network regulated by p14ARF. For both tests it is possible to observe overexpressed Mdm2-mediated p53 degradation as carcinogenic initial behavior at first 24 hours. In **(A)**, with control action with $(k_a u_1)$ in a range between 0 between 1.5 proteins/s, increased expression of p53 response is not achieved (24–48 hours), and likewise in **(B)** increased expression of p53 response under pinning control (24–72 hours) with $(k_a u_1)$ in a range between 9.6 and 50 proteins/s is achieved.

The increased expression of p53 levels by means of the pinning control technique is achieved as shown in Fig. 7.8. Such technique generates the desired behavior of p53 progressive accumulation, assuming that this behavior has a post-translational activation mediated by ATM. This could generate the activation of downstream proteins, contributing to the activation of apoptosis or cell cycle arrest [13, 43, 14, 45]. The pinning control (7.10) is located at the production of p14ARF node with $K_i = 5$, which yields an increased p53 level expression and guarantees that p14ARF production $k_a u_1$ achieves a range between 9.6 and 50 proteins/s. p14ARF is moved from the nucleolus to nucleoplasm in response to DNA damage, where $Mdm2_{nuclear}$-p14ARF complex promotes p53 tumor suppressor activity. For this case, it is possible to generate p53 accumulation, which is assumed to be activated by a mechanism linked to post-radiation activation of the ATM protein and p53-dependent induction of downstream proteins [3, 42, 31]. With the pinning control law, it is possible to produce scenarios with different physiological or pathological responses of the p53-Mdm2 network .

For both cases, the control law $(u_1(t) \in \mathbb{R})$ is applied to p14ARF node as in equation (7.13). The respective control actions are displayed in Fig. 7.9(A) for the first case and Fig. 7.9(B) for the second case. From the above results, it can be clearly seen that the pinning controller achieves regulation successfully for the p53-Mdm2 network .

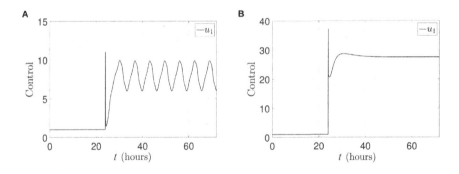

FIGURE 7.9

Control signal $u_1(t)$ applied in p14ARF. (**A**) Case 1: control action to achieve an oscillatory pattern under gamma-radiation. (**B**) Case 2: control action to achieve an increased expression of p53 response.

7.6   Conclusions

In this chapter, the two cases discussed above, the pinning control strategy
for the p53-Mdm2 network dynamics is applied on the p14ARF node based
on sensitivity analysis as follows: In the first case, oscillatory pattern activity
is achieved; in the second case, p53 increased expression and accumulation
are obtained. By means of sensitivity analysis in K_a with respect to $p53$ and
$Mdm2_{nuclear}$, it is shown that with K_a low values, the network does not
reach the desired behavior. However, for the adequate K_a values mentioned
above, the network recovers its oscillatory pattern behavior, or an increase
in p14ARF production leads to a consistently increase p53 level. The
pinning control strategy suppresses Mdm2 prooncogenic behavior and allows
functional recovery of p53 physiological response.

Bibliography

[1] M. Ashcroft, Y. Taya, and K. Vousden. Stress signals utilize multiple pathways to stabilize p53. *Molecular and Cellular Biology*, 20(9):3224–3233, 2000.

[2] A. Arya, A. El-Fert, T. Devling, R. Eccles, M. Aslam, C. Rubbi, N. Vlatković, J. Fenwick, et al. Nutlin-3, the small-molecule inhibitor of MDM2, promotes senescence and radiosensitises laryngeal carcinoma cells harbouring wild-type p53. *British Journal of Cancer*, 103(2):186, 2010.

[3] C. Bakkenist and M. Kastan. DNA damage activates ATM through intermolecular autophosphorylation and dimer dissociation. *Nature*, 421(6922):499, 2003.

[4] E. Batchelor, A. Loewer, C. Mock, and G. Lahav. Stimulus-dependent dynamics of p53 in single cells. *Molecular Systems Biology*, 7(1):488, 2011.

[5] G. Bond, W. Hu, E. Bond, H. Robins, S. Lutzker, A. Stuart, N. Arva, et al. A single nucleotide polymorphism in the MDM2 promoter attenuates the p53 tumor suppressor pathway and accelerates tumor formation in humans. *Cell*, 119(5):591–602, 2004.

[6] C. Camus, M. Higgins, D. Lane, and S. Lain. Differences in the ubiquitination of p53 by Mdm2 and the HPV protein E6. *FEBS Letters*, 536(1-3):220–224, 2003.

[7] X. Chen, L. Ko, L. Jayaraman, and C. Prives. p53 levels, functional domains, and DNA damage determine the extent of the apoptotic response of tumor cells. *Genes & Development*, 10(19):2438–2451, 1996.

[8] L. Chen, S. Willis, A. Wei, B. Smith, J. Fletcher, et al. Differential targeting of prosurvival Bcl-2 proteins by their BH3-only ligands allows complementary apoptotic function. *Molecular Cell*, 17(3):393–403, 2005.

[9] G. Chen. Pinning control and controllability of complex dynamical networks. *International Journal of Automation and Computing*, 1–9, 2017.

[10] A. Ciliberto, B. Novák, and J. Tyson. Steady states and oscillations in the p53/Mdm2 network. *Cell Cycle*, 4(3):488–493, 2005.

[11] A. Dei Tos, C. Doglioni, S. Piccinin, R. Sciot, A. Furlanetto, et al. Coordinated expression and amplification of the MDM2, CDK4, and HMGI-C genes in atypical lipomatous tumours. *The Journal of Pathology*, 190(5):531–536, 2000.

[12] R. Dickinson and R. Gelinas. Sensitivity analysis of ordinary differential equation systems: A direct method. *Journal of Computational Physics*, 21(2):123–143, 1976.

[13] W. El-Deiry. Regulation of p53 downstream genes. *Seminars in Cancer Biology*, 8(5):345–357, 1998.

[14] N. Geva-Zatorsky, N. Rosenfeld, S. Itzkovitz, et al. Oscillations and variability in the p53 system. *Molecular Systems Biology*, 2(1):1–13, 2006.

[15] D. Green and G. Kroemer. The pathophysiology of mitochondrial cell death. *Science*, 305(5684):626–629, 2004.

[16] A. Hafner, J. Stewart-Ornstein, J. Purvis, W. Forrester, M. Bulyk, and G. Lahav. p53 pulses lead to distinct patterns of gene expression albeit

similar DNA-binding dynamics *Nature Structural and Molecular Biology*, 24(10):840, 2017.

[17] R. Hage-Sleiman, H. Bahmad, H. Kobeissy, Z. Dakdouk, F. Kobeissy, and G. Dbaibo. Genomic alterations during p53-dependent apoptosis induced by γ-irradiation of Molt-4 leukemia cells. *PloS One*, 12(12):e0190221, 2017.

[18] D. Hamby. A review of techniques for parameter sensitivity analysis of environmental models. *Environmental Monitoring and Assessment*, 32(2):135–154, 1994.

[19] A. Hsing, D. Faller, and C. Vaziri. DNA-damaging aryl hydrocarbons induce Mdm2 expression via p53-independent post-transcriptional mechanisms. *Journal of Biological Chemistry*, 275(34):26024–26031, 2000.

[20] S. Khan, C. Guevara, G. Fujii, and D. Parry. p14ARF is a component of the p53 response following ionizing irradiation of normal human fibroblasts. *Oncogene*, 23(36):6040–6046, 2004.

[21] G. Lahav, N. Rosenfeld, A. Sigal, N. Geva-Zatorsky, A. Levine, M. Elowitz, and U. Alon. Dynamics of the p53-Mdm2 feedback loop in individual cells. *Nature Genetics*, 36(2):147–150, 2004.

[22] D. Lane. p53, guardian of the genome. *Nature*, 358:15–16, 1992.

[23] G. Leenders and J. Tuszynski. Stochastic and deterministic models of cellular p53 regulation. *Frontiers in Oncology*, 3:64, 2013.

[24] F. Lewis, H. Zhang, K. Hengster-Movric, and A. Das. *Cooperative Control of Multi-agent Systems: Optimal and Adaptive Design Approaches*. London, UK: Springer-Verlag, 2013.

[25] X. Li, X. Wang, and G. Chen. Pinning a complex dynamical network to its equilibrium. *IEEE Transactions on Circuits and Systems I: Regular Papers*, 51(10):2074–2087, 2004.

[26] J. Momand, G. Zambetti, D. Olson, D. George, and A. Levine. The mdm-2 oncogene product forms a complex with the p53 protein and inhibits p53-mediated transactivation. *Cell*, 69(7):1237–1245, 1992.

[27] M. Nilbert, F. Mitelman, N. Mandahl, A. Rydholm, and H. Willén. MDM2 gene amplification correlates with ring chromosomes in soft tissue tumors. *Genes, Chromosomes and Cancer*, 9(4):261–265, 1994.

[28] J. Oliner, K. Kinzler, P. Meltzer, D. George, and B. Vogelstein. Amplification of a gene encoding a p53-associated protein in human sarcomas. *Nature*, 358(6381):80, 1992.

[29] J. Oliner, J. Pietenpol, S. Thiagalingam, J. Gyuris, K. Kinzler, and B. Vogelstein. Oncoprotein MDM2 conceals the activation domain of tumour suppressor p53. *Nature*, 362(6423):857, 1993.

[30] T. Parisi, A. Pollice, A. Di Cristofano, V. Calabrò and G. La Mantia. Transcriptional regulation of the human tumor suppressor p14ARF by E2F1, E2F2, E2F3, and Sp1-like factors. *Biochemical and Biophysical Research Communications*, 291(5):1138–1145, 2002.

[31] S. Pauklin, A. Kristjuhan, T. Maimets, V. Jaks. ARF and ATM/ATR cooperate in p53-mediated apoptosis upon oncogenic stress. *Biochemical and Biophysical Research Communications*, 334(2):386–394, 2005.

[32] J. Purvis, K. Karhohs, C. Mock, E. Batchelor, A. Loewer, and G. Lahav. p53 dynamics control cell fate. *Science*, 336(6087):1440–1444, 2012.

[33] E. Rayburn, R. Zhang, J. He, and .H. Wang. MDM2 and human malignancies: expression, clinical pathology, prognostic markers, and

implications for chemotherapy. *Current Cancer Drug Targets*, 5(1):27–41, 2005.

[34] W. Ren and R. Beard. *Distributed Consensus in Multi-vehicle Cooperative Control.* London, UK: Springer-Verlag, 2008.

[35] T. Riley, E. Sontag, P. Chen,and A. Levine, Arnold. Transcriptional control of human p53-regulated genes. *Nature Reviews Molecular Cell Biology*, 9(5):402, 2008.

[36] M. Scheffner, B. Werness, J. Huibregtse, A. Levine, and P. Howley. The E6 oncoprotein encoded by human papillomavirus types 16 and 18 promotes the degradation of p53. *Cell*, 63(6):1129–1136, 1990.

[37] R. Sionov and Y. Haupt. The cellular response to p53: The decision between life and death. *Oncogene*, 18(45):6145, 1999.

[38] J. Slotine and W. Li. *Applied Nonlinear Control.* Englewood Cliffs, NJ, USA: Prentice Hall, 1991.

[39] L. Strigari, M. Mancuso, V. Ubertini, A. Soriani, P. Giardullo, et al. Abscopal effect of radiation therapy: Interplay between radiation dose and p53 status. *International Journal of Radiation Biology*, 90(3):248–255, 2014.

[40] O. Suarez, C. Vega, E. Sanchez, A. González, O. Rodríguez, and A. Pardo. Degradación anormal de p53 e inducción de apoptosis en la red p53-Mdm2 usando la estrategia de control tipo pin. *Revista Colombiana de Tecnologías de Avanzada*, 2(32):1–7, 2018.

[41] S. Sykes, H. Mellert, M. Holbert, K. Li, et al. Acetylation of the p53 DNA-binding domain regulates apoptosis induction. *Molecular Cell*, 24(6):841–851, 2006.

[42] A. Villunger, E. Michalak, L. Coultas, F. Müllauer, G. Böck, et al. p53-and drug-induced apoptotic responses mediated by BH3-only proteins puma and noxa. *Science*, 302(5647):1036–1038, 2003.

[43] J. Wagner, L. Ma, J. Rice, W. Hu, A. Levine, and G. Stolovitzky. p53–Mdm2 loop controlled by a balance of its feedback strength and effective dampening using ATM and delayed feedback. *IEE Proceedings-Systems Biology*, 152(3):109–118, 2005.

[44] X. Wang and G. Cheng. Pinning control of scale-free dynamical networks. *Physica A: Statistical Mechanics and its Applications*, 310(3):521–531, 2002.

[45] K. Wee, U. Surana, and B. Aguda. Oscillations of the p53-Akt network: Implications on cell survival and death. *PloS One*, 4(2):e4407, 2009.

[46] K. Wee, W. Yio, U. Surana, and K. Chiam. Transcription factor oscillations induce differential gene expressions. *Biophysical Journal*, 102(11):2413–2423, 2012.

[47] R. Weinberg, D. Veprintsev, M. Bycroft, and A. Fersht. Transcription factor oscillations induce differential gene expressions. *Biophysical Journal*, 102(11):2413–2423, 2012.

[48] Y. Zhang, Y. Xiong, and W. Yarbrough. ARF promotes MDM2 degradation and stabilizes p53: ARF-INK4a locus deletion impairs both the Rb and p53 tumor suppression pathways. *Cell*, 92(6):725–734, 1998.

8

Secondary Control of Microgrids

In this chapter, a new control scheme of secondary voltage and frequency control based on a discrete-time neural inverse optimal distributed cooperative structure is introduced for islanded microgrids. A neural adaptive secondary controller for each distributed generator (DG) is developed to achieve the desired goals. The designed controllers are composed of an on-line neural identification scheme on the basis of an RHONN using the EKF and a nonlinear control strategy. Additionally, the control scheme does not require information of all installed distributed generators, nor a distributed generator model, which improves reliability. The controllers are verified through simulation for an islanded AC microgrid.

8.1 Microgrid Control Structure

Fig. 8.1 displays the considered hierarchical control structure for an AC microgrid to ensure adequate operation in islanded mode [3]. The designed control scheme is composed of an inner-loop controller (primary control) and an outer-loop one (secondary control); both controllers are applied for the voltage source inverters (VSI) of DGs.

The primary controller regulates voltage and frequency of the microgrid for the operating modes (grid-connected mode and islanded mode). The goal

Secondary Frequency Control

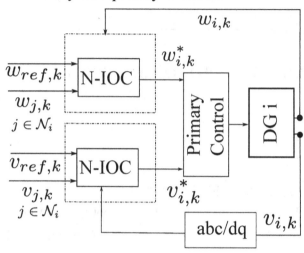

FIGURE 8.1
Block diagram of the distributed cooperative hierarchical control of a microgrid.

of this layer is to prevent voltage and frequency failures [5]. Nevertheless, this controller is insufficient to ensure normal operating conditions of the microgrid, requiring an additional control level to compensate voltage and frequency deviations. The secondary control level responds in a slower time-scale, which allows decoupling between the primary and the secondary control layers and simplifies their designs. Generally, the control strategies of the secondary controllers are VF control, PQ control, and droop control [4]. In this chapter, the designed scheme for the secondary control level is voltage-frequency neural inverse optimal distributed cooperative control. The secondary controller for each DG requires minimal information from neighbors. This control adapts to changes in the DG parameters, loads, and microgrid operating conditions. The controller reacts to changes in the system, adjusting the control parameters in real-time, thanks to the on-line neural identifier.

The primary control is usually implemented as a local controller at each DG. Development of the primary local controller is conventionally based on active and reactive power droop techniques [8]. Droop technique relates the frequency and active power, and voltage amplitude and reactive power [9]. Frequency and voltage droop characteristics for the ith DG are given by

$$w_{i,k} = w^*_{i,k} - D_P P_{i,k}, \qquad (8.1)$$

$$v_{o,mag,k} = v^*_{i,k} - D_Q Q_{i,k}. \qquad (8.2)$$

The droop technique gives voltage references for the voltage control loop, implemented as a discrete-time proportional-integral (PI) controller, which adjusts current references for the current control loop. This latter loop manipulates the Sinusoidal Pulse Width Modulation (SPWM) algorithm for the inverter bridge, which is connected to a primary DC power source. The current control loop is designed as a discrete-time PI controller with feedforward signal and harmonics rejection filter.

8.2 Distributed Cooperative Secondary Control

The objective of this controller is to determine the references trajectory for the primary control ($w^*_{i,k}$, $v^*_{i,k}$) to regulate frequency and voltage amplitude to their nominal values. For this goal, a neural adaptive and distributed secondary control for microgrids with inverter-based DG is designed. The DG nonlinear dynamical model and parameters are assumed to be unknown.

8.2.1 Secondary frequency control

The secondary frequency control of the ith DG selects $w^*_{i,k}$ in (8.1), such that frequency amplitude $w_{i,k}$ of each DG approaches the reference value, w_{ref}.

Assume that there is an unknown discrete-time dynamical model for $w_{i,k} = f(w_{i,k}^*)$, which is formulated as a second-order nonlinear dynamical system, given by

$$x_{Wi,k+1} = f_{Wi}(x_{Wi,k}) + g_{Wi}(x_{Wi,k})u_{Wi,k},$$

$$y_{Wi,k} = w_{i,k}, \quad \text{with} \quad u_{Wi,k} = w_{i,k}^*,$$

where the state vector for each ith DG inverter is $x_{Wi,k} = [w_{i,k} \quad w_{i,k+1}]^T$, the vector functions $f_{Wi}(\cdot)$ and $g_{Wi}(\cdot)$ are unknown, with $x_{Wi,k}$ available for measurement.

The DGs are considered as nodes, which exchange information among neighbors through a communication network. This communication network is modeled by a graph [2]. The local neighborhood tracking error for each DG is defined by

$$e_{Wi,k} = \sum_{j \in \mathcal{N}_i} a_{ij}(y_{Wi,k} - y_{Wj,k}) + b_i(y_{Wi,k} - y_{W0,k}), \qquad (8.3)$$

where $\mathcal{N}_i$ denotes the set of DGs neighboring the ith DG, a_{ij} denotes the elements of the communication adjacency matrix, $y_{W0,k} = [w_{ref,k} \quad w_{ref,k+1}]^T$, and b_i is the pinning gain by which the ith DG is connected to the leader node, the reference one.

The dynamical equation of (8.3) is

$$e_{Wi,k+1} = \sum_{j \in \mathcal{N}_i} a_{ij}(y_{Wi,k+1} - y_{Wj,k+1}) + b_i(y_{Wi,k+1} - y_{W0,k+1}) \qquad (8.4)$$

$$= d_i(f_{Wi}(x_{Wi,k}) + g_{Wi}(x_{Wi,k})u_{Wi,k}) - b_i y_{W0,k+1}$$

$$+ \sum_{j \in \mathcal{N}_i} a_{ij}(f_{Wj}(x_{Wj,k}) + g_{Wj}(x_{Wj,k})u_{Wj,k})$$

$$\text{with } d_i = b_i + \sum_{j \in \mathcal{N}_i} a_{ij}$$

$$e_{Wi,k+1} = \tilde{F}_{Wi}(x_{Wi,k}) + \tilde{G}_{Wi}(x_{Wi,k})u_{Wi,k} + \sum_{j \in \mathcal{N}_i} a_{ij}H_j(y_{Wj,k})$$

where the functions $\tilde{F}_{Wi}(\cdot)$, $\tilde{G}_{Wi}(\cdot)$, and $H_j(\cdot)$ are unknown. System (8.4) is identified by an RHONN identifier in a series-parallel structure, which is

$$\chi_{Wi,k+1} = \hat{F}_{Wi}(e_{Wi,k}) + \hat{G}_{Wi} u_{Wi,k} \qquad (8.5)$$

where χ_{Wi} is the vector estimated by the neural identifier, $\hat{G}_{Wi} = [0,\ \varpi_{Wi,2}]^T$, and

$$\hat{F}_{Wi}(e_{Wi,k}) = \left[\hat{f}_{Wi,1}(e_{Wi,k}),\ \hat{f}_{Wi,2}(e_{Wi,k}) \right]^T$$

is defined as

$$\hat{f}_{Wi,1}(e_{Wi,k}) = \omega_{1,k} S(e_{Wi,1,k}) + e_{Wi,2,k}$$
$$= \omega_{11,k} S(e_{Wi,1,k}) + \omega_{12,k} S^2(e_{wi,1,k}) + e_{Wi,2,k}$$
$$\hat{f}_{Wi,2}(e_{Wi,k}) = \omega_{21,k} S(e_{Wi,2,k}) + \omega_{22,k} S(e_{Wi,1,k}) S(e_{Wi,2,k})$$
$$+ \omega_{23,k} S^2(e_{Wi,2,k}) + \omega_{24,k} S^3(e_{Wi,2,k})$$

with $i = 1, \ldots, n$ is the DG numbers. For on-line RHONN weights adaptation, the EKF training algorithm is used [6].

System (8.5) is an affine-in-the-input discrete-time nonlinear system. Furthermore, the discrete-time inverse optimal controller is designed [7] to stabilize (8.4), as

$$z_{Wi,1,k} = e_{Wi,1,k}$$
$$V_{Wi,2,k} = -\omega_{1,k} S(e_{Wi,1,k}) - K_{Wi,1} z_{Wi,1,k}$$
$$u^*_{Wi,k} = -\frac{1}{2}\left(R + \frac{g_k^T P g_k}{2} \right)^{-1} g_k^T P (f_k - x_{d,k+1})$$
$$= -\frac{1}{3}(\hat{f}_{Wi,2}(e_{Wi,k}) - V_{Wi,2,k+1})$$

where $z_{Wi,1,k}$ is an auxiliary variable, $V_{Wi,2,k}$ is a virtual control, $0 < K_{Wi,1} <$

1 is a parameter design, and $u^*_{Wi,k}$ is a discrete-time inverse optimal control law, with $P = 1$, $R = 1$, $g_k = \varpi_{Wi,2} = 1$.

8.2.2 Secondary voltage control

The secondary voltage control of the ith DG in the microgrid selects $v^*_{i,k}$ in (8.2), such that the voltage magnitude $v_{o,mag,k}$ of each DG approaches the reference value, v_{ref}. The voltage magnitude for each ith DG is

$$v_{o,mag,k} = \sqrt{v^2_{d,i,k} + v^2_{q,i,k}},$$

and the primary voltage control aligns the voltage magnitude on the d-axis of the corresponding reference frame ($v^2_{q,i,k} = 0$); furthermore, the secondary control goal is to design $v^*_{i,d,k}$ such that $v_{i,d,k} \to v_{ref}$, for all i [1].

For secondary frequency control, assume that there is an unknown discrete-time dynamical model $v_{i,d,k} = f(v^*_{i,d,k})$, which is formulated as a second-order nonlinear dynamical system, given by

$$x_{Vi,k+1} = f_{Vi}(x_{Vi,k}) + g_{Vi}(x_{Vi,k})u_{Vi,k},$$

$$y_{Vi,k} = v_{i,d,k}, \quad \text{with} \quad u_{Vi,k} = v^*_{i,d,k},$$

where the state vector for each ith DG inverter is $x_{Vi,k} = [v_{i,d,k}, \quad v_{i,d,k+1}]^T$, the vector functions $f_{Vi}(\cdot)$ and $g_{Vi}(\cdot)$ are unknown, with $x_{Vi,k}$ available for measurement.

For this case, the local neighborhood tracking error for each DG is defined by

$$e_{Vi,k} = \sum_{j \in \mathcal{N}_i} a_{ij}(y_{Vi,k} - y_{Vj,k}) + b_i(y_{Vi,k} - y_{V_0,k}), \qquad (8.6)$$

where $y_{V_0,k} = [v_{ref,k}, \quad v_{ref,k+1}]^T$.

The dynamical equation of (8.6) is

$$e_{Vi,k+1} = \sum_{j \in \mathcal{N}_i} a_{ij}(y_{Vi,k+1} - y_{Vj,k+1}) + b_i(y_{Vi,k+1} - y_{V0,k+1}) \tag{8.7}$$

$$= \tilde{F}_{Vi}(x_{Vi,k}) + \tilde{G}_{Vi}(x_{Vi,k})u_{Vi,k} + \sum_{j \in \mathcal{N}_i} a_{ij} H_j(y_{Wj,k})$$

where the functions $\tilde{F}_{Wi}(\cdot)$ $\tilde{G}_{Wi}(\cdot)$, and $H_j(\cdot)$ are unknown. System (8.7) is identified by an RHONN identifier in a series-parallel structure, which is

$$\chi_{Vi,k+1} = \hat{F}_{Vi}(e_{Vi,k}) + \hat{G}_{Vi} u_{Vi,k} \tag{8.8}$$

where χ_{Vi} is the vector estimated by the neural identifier, $\hat{G}_{Vi} = \hat{G}_{Wi}$, and

$$\hat{F}_{Vi}(e_{Vi,k}) = \hat{F}_{Wi}(e_{Wi,k})$$

The discrete-time inverse optimal controller is designed [7] to stabilize (8.7), as

$$z_{Vi,1,k} = e_{Vi,1,k}$$
$$V_{Vi,2,k} = -\omega_{1,k} S(e_{Vi,1,k}) - K_{Vi,1} z_{Vi,1,k}$$
$$u^*_{Vi,k} = -\frac{1}{3}(\hat{f}_{Vi,2}(e_{Vi,k}) - V_{Vi,2,k+1})$$

where $z_{Vi,1,k}$ is an auxiliary variable, $V_{Vi,2,k}$ is a virtual control, $0 < K_{Vi,1} < 1$ is a parameter design, and $u^*_{Vi,k}$ is a discrete-time inverse optimal control law, with $P = 1$, $R = 1$, $g_k = \varpi_{Wi,2} = 1$.

8.3   Simulation Results

The designed secondary voltage and frequency controllers, as well as the microgrid, are simulated using the SimPower System toolbox of Matlab. The effectiveness of the control scheme is verified by simulating a microgrid in island operation mode composed of four DGs, as shown in Fig. 8.2.

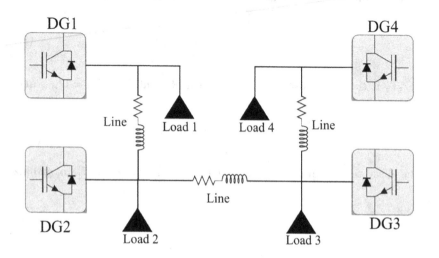

FIGURE 8.2
Microgrid selected structure.

The parameters of the proposed microgrid are illustrated in Table 8.1.

TABLE 8.1
The microgrid equipment and parameters.

DERs	Equipment	Parameters
DG 1 to DG4	DG: S, V_{dc}, F	100 kVA, 500 V, 60 Hz
	LC filter: R, L, C	0.03 Ω, 0.35 mH, 50 μF
Line 1 to 3	R, L	0.23 Ω, 0.318 mH
Loads 1 to 4	P_L, Q_L	100 kW, 24 kVAr

The designed secondary controller is used to deal with grid voltage and frequency deviations. In the case of microgrid voltage amplitude and frequency disturbances, this controller sets the desired values of the active and reactive power, which should be injected by each DG into the microgrid. The selected power values are tracked using the primary DG controller. Since the control scheme is based on pinning control, all nodes (DGs) are synchronized to the leader one, which in the present case is DG 1. This leader node shares the information through a communication network with the other DGs (DG2-DG4); hence, all the installed DGS references are synchronized.

Fig. 8.3 displays the dynamics of the $d-q$ voltage in (pu) for each DG. The $-d-$ lines voltage (-.) of each DG is forced to track constant amplitude $1pu$ (-), while the $-q-$ lines voltage $(--)$ is maintained at zero. Fig. 8.4 presents

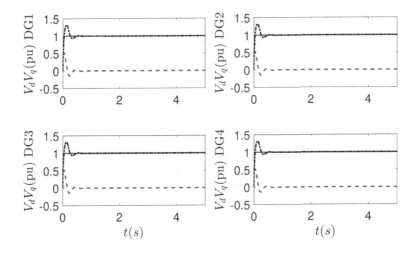

FIGURE 8.3
The $d-q$ voltage and reference, DGi, $i = 1, \ldots, 4$: $v_{d,i,k}$ (-. line), $v_{q,i,k}$ $(--$ line), v_{ref} (- line).

the frequency desired values (-line) and the real one at each DG (-. line). The active (-.line) and reactive $(--$line) power desired values of each DG, which are obtained the secondary control scheme, are presented at Fig. 8.5. These desired power trajectories are tracked using the primary controllers. Since the microgrid is simulated considering balanced loads conditions, the

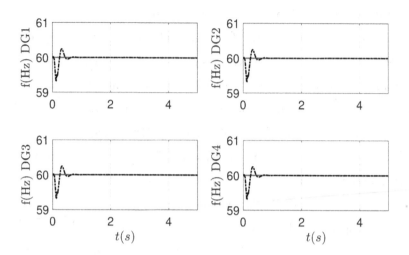

FIGURE 8.4

Frequency DGi, $i = 1, \ldots, 4$: $w_{i,k}$ (-. line), w_{ref} (- line).

active power desired value is $1pu$, ensuring a nominal frequency value of $60Hz$, and the reactive power desired value is zero, assuring a microgrid nominal voltage amplitude value. Figs. 8.6 and 8.7 illustrate the three-phase voltage

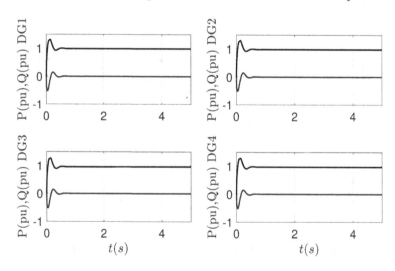

FIGURE 8.5

Active and reactive power DGi, $i = 1, \ldots, 4$: $P_{i,k}$ (-.line), $Q_{i,k}$ ($-$ $-$line).

and currents lines for each DG (*pu*), respectively. From the obtained results,

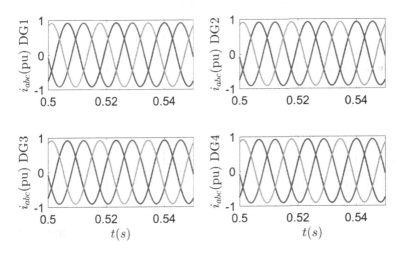

FIGURE 8.6
Three-phase currents, DGi, $i = 1, \ldots, 4$.

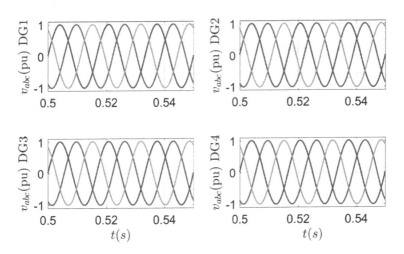

FIGURE 8.7
Three-phase voltages DGi, $i = 1, \ldots, 4$.

it is clear that the designed control scheme achieves the control objectives, where the rated voltage amplitude and frequency are ensured. In addition, an adequate decoupling between the $d - q$ voltage axes, and active and reactive power, are carried-out, which guarantee a flexible power-sharing.

8.4   Conclusions

Secondary voltage and frequency controllers, based on a discrete-time neural inverse optimal distributed cooperative scheme, are developed for islanded AC microgrids. The designed controllers are used to generate reference trajectories for the primary level control. Since they are based on RHONN identifiers, trained on-line with an EKF, which allows adjusting indirectly the control algorithm in real-time, DGs mathematical models are not required. Additionally, thanks to the pinning control technique, all nodes (DGs) are synchronized to the leader node, which generates the reference, and is connected to a minimal number of DGs; each node shares the information with its neighbors through a simple communication network. Simulation results illustrate the effectiveness of the control scheme to ensure maintaining the required voltage amplitude and frequency in the microgrid. In addition, adequate active and reactive power sharing is achieved.

Bibliography

[1] A. Bidram, A. Davoudi, F. L. Lewis, and Z. Qu. Secondary control of microgrids based on distributed cooperative control of multi-agent systems. *IET Generation, Transmission & Distribution*, 7(8): 822–831, 2013.

[2] A. Bidram, A. Davoudi, F. L. Lewis, and J. M. Guerrero. Distributed cooperative secondary control of microgrids using feedback linearization. *IEEE Transactions on Power Systems*, 28(3): 3462–3470, 2013.

[3] H. Han, Y. Liu, Y. Sun, M. Su, and J. M. Guerrero. An improved droop control strategy for reactive power sharing in islanded microgrid. *IEEE Transactions on Power Electronics*, 30(6): 3133–3141, 2014.

[4] J.-Y. Kim, J.-H. Jeon, S.-K. Kim, C. Cho, J. H. Park, H.-M. Kim, and K.-Y. Nam. Cooperative control strategy of energy storage system and microsources for stabilizing the microgrid during islanded operation. *IEEE Transactions on Power Electronics*, 25(12): 3037–3048, 2010.

[5] N. Pogaku, M. Prodanovic, and T. C. Green. Modeling, analysis and testing of autonomous operation of an inverter-based microgrid. *IEEE Transactions on Power Electronics*, 22(2): 613–625, 2007.

[6] E. N. Sanchez, A. Y. Alanís, and A. G. Loukianov. *Discrete-Time High-Order Neural Control: Trained with Kalman Filtering*. Berlin, Germany: Springer, 2008.

[7] E. N. Sanchez and F. Ornelas-Tellez. *Discrete-Time Inverse Optimal Control for Nonlinear Systems*. Boca Raton, FL, USA: CRC Press, 2017.

[8] Y. Sun, X. Hou, J. Yang, H. Han, M. Su, and J. M. Guerrero. New perspectives on droop control in AC microgrid. *IEEE Transactions on Industrial Electronics*, 64(7): 5741–5745, 2017.

[9] U. B. Tayab, M. A. Roslan, L. J. Hwai, and M. Kashif. A review of droop control techniques for microgrid. *Renewable and Sustainable Energy Reviews*, 76: 717–727, 2017.

Index

Printed in the United States
by Baker & Taylor Publisher Services